Luciano Floridi
Editor

# Protection of Information and the Right to Privacy – A New Equilibrium?

*Editor*
Luciano Floridi
Oxford Internet Institute
University of Oxford
Oxford
UK

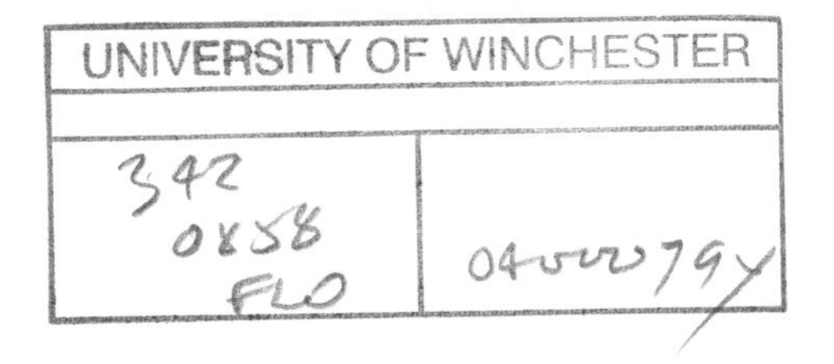


ISSN 2352-1902 ISSN 2352-1910 (electronic)
ISBN 978-3-319-05719-4 ISBN 978-3-319-05720-0 (eBook)
DOI 10.1007/978-3-319-05720-0
Springer Cham Heidelberg New York Dordrecht London

Library of Congress Control Number: 2014938361



Printed on acid-free paper

Springer is part of Springer Science+Business Media (www.springer.com)

# Protection of Information and the Right to Privacy – A New Equilibrium?

# Law, Governance and Technology Series

VOLUME 17

*Series Editors:*

POMPEU CASANOVAS, *Institute of Law and Technology, UAB, Spain*

GIOVANNI SARTOR, *University of Bologna (Faculty of Law – CIRSFID) and European, University Institute of Florence, Italy*



For further volumes:
http://www.springer.com/series/8808

# Preface

This volume collects the fully revised texts of the papers presented at the international workshop on "Protection of Information and the Right to Privacy: A New Equilibrium?" held at the European University Institute, Fiesole, 21st June 2013. The meeting was organised by the UNESCO Chair in Information and Computer Ethics, in collaboration with Google and the Department of Law, and the Global Governance Program at the European University Institute.

The idea of the workshop was to deal head-on with the difficult but necessary balance that liberal societies must find between freedom of information (understood as freedom of speech and as the opposite of censorship), security, and privacy. Let me explain.

Communication means exchanging messages. So even the most elementary act of communication involves four elements: a sender, a receiver, a message, and a referent of the message. All this is unproblematic. What may cause difficulties is the fact that, at different times in human history, these components have been associated with a variety of rights and freedoms. *Freedom of speech* concerns the right to send messages. *Freedom of information*, understood as the opposite of censorship, concerns the right to receive messages. *Communication security* concerns the right to see the message protected from unwanted intrusion. And the right to *informational privacy* is the freedom from being the referent of a message. Finding an equilibrium in this square has always been a matter of delicate negotiation. However, it was a less daunting task when few identifiable senders sent easily accessable messages to few identifiable receivers, about clearly checkable referents, with very limited tools to undermine the safety of the message. Digital Information and Communication Technologies (ICTs) have disrupted all this irreversibly. Today, the number and kinds of senders and receivers have grown exponentially, to become virtually limitless; the quantities and types of messages are already staggering; the variety and sophistication of malicious applications are a growing threat; and the nature and scope of the referents is now potentially boundless. Phenomena such as citizen journalism, once inconceivable, have become commonplace. Governments and companies produce and deploy surveillance technologies on a vast scale. Indeed, we are discovering that the old equilibrium was achievable also thanks to constraints that ICTs have either removed or are increasingly eroding.

The result is that ICTs are redesigning the equilibrium between freedom of speech, freedom of information, information security, and the right to informational privacy. New tensions and potential incompatibilities within this square keep emerging with increasing pressure. Clearly, finding a new balance has become a pressing task in any information society that seeks to implement informational rights and freedoms while fostering technological innovation, higher standards of living, and human well-being.

In this complex scenario, it seems that old legal and ethical frameworks may need to be not only updated, but also supplemented and complemented by new conceptual solutions. Neither a conservative attitude ("more of the same") nor a revolutionary zeal ("never seen before") is likely to lead to satisfactory solutions. Instead, more reflection and better conceptual design are needed, not least to harmonise different perspectives and legal frameworks across countries. So the task of the workshop was to contribute to fill the serious gap identified above. In particular, the focus was on how we may reconcile high levels of information security with robust degrees of informational privacy, thus solving the new tensions and build a fair, shareable, and sustainable balance. Of course, this is not an easy task. But our hope is that the workshop and this volume may contribute to identifying solutions, resolving problems, and anticipating difficulties in such a vital area of human interactions.

The format of the workshop was that of an invitation-only, one-day meeting. Only a selected number of experts and members of the audience was invited to participate, so it was possible to take for granted most background knowledge and focus immediately and sharply on the issues that we considered more pressing. Pre-meeting drafts of the papers were made available to all participants, so that we could all become acquainted with the intellectual agenda before coming together for the meeting. This is also why it was possible to publish this volume within a rather reasonable timespan.

The workshop saw the participation of a selection of distinguished experts, here listed in order of presentation, with the updated titles provided for the new chapters of the volume: Giovanni Sartor (University of Bologna & European University Institute), *The right to be forgotten: dynamics of privacy and publicity*, Ugo Pagallo and Massimo Durante (University of Turin), *Legal memories and the right to be forgotten*; Mireille Hildebrandt (Radboud University Nijmegen), *Location Data, Purpose Binding and Contextual Integrity: What's the Message?*; Dawn Nunziato (George Washington University), *With Great Power Comes Great Responsibility: Proposed Principles of Digital Due Process for ICT Companies*; Hosuk Lee-Makiyama (European Centre for International Political Economy (ECIPE), *The Political Economy of Data: EU Privacy Regulation and the International Redistribution of Its Costs*. My chapter, entitled *The Rise of the MASs*, closes the volume.

I shall not summarise the contents of the chapters here, since a short abstract introduces each of them. What I may mention, by way of conclusion, is that, with a bit of luck, if it can be so-called, the "Prism scandal" hit the news just a few days before the workshop took place,[1] proving, if still necessary, that the aforementioned equilibrium is crucial in democratic, information societies.

— L. Floridi

[1] Gellman, Barton; Poitras, Laura (June 6, 2013). "US Intelligence Mining Data from Nine U.S. Internet Companies in Broad Secret Program". The Washington Post. Greenwald, Glenn; MacAskill, Ewen (June 6, 2013). "NSA Taps in to Internet Giants' Systems to Mine User Data, Secret Files Reveal—Top-Secret Prism Program Claims Direct Access to Servers of Firms Including Google, Apple and Facebook—Companies Deny Any Knowledge of Program in Operation Since 2007—Obama Orders US to Draw Up Overseas Target List for Cyber-Attacks". The Guardian.

# Acknowledgements

I am very grateful to all the speakers for their presentations and then for their chapters and to the invited participants for their feedback. The meeting and this book would not have been possible without the essential support and help, at different stages, of Laura Bononcini, Penny Driscoll, William Echikson, Rosanna Lewis, Ugo Pagallo, Marco Pancini, and Giovanni Sartor.

# Contents

# Contributors

**Massimo Durante** Law School, University of Turin, Torino, Italy

**Luciano Floridi** Oxford Internet Institute, University of Oxford, Oxford, UK

**Mireille Hildebrandt** Institute of Computing and Information Sciences (iCIS), Radboud University, Nijmegen, The Netherlands

Jurisprudence, Erasmus School of Law (ELS), Rotterdam, The Netherlands

Law Science Technology & Society studies (LSTS), Vrije Universiteit, Brussels, The Netherlands

**Hosuk Lee-Makiyama** European Centre for International Political Economy (ECIPE), Brussels, Belgium

**Dawn Carla Nunziato** George Washington University Law School, Washington, DC, USA

**Ugo Pagallo** Law School, University of Turin, Torino, Italy

**Giovanni Sartor** Legal Informatics, University of Bologna, Bologna, Italy

Legal informatics and Legal Theory, European University Institute of Florence, Florence, Italy

**Mariarosaria Taddeo** Political & International Studies, University of Warwick, Coventry, UK

University of Oxford, Oxford, UK

# About the Authors

**Massimo Durante** is Researcher in Philosophy of Law at the Department of Law of the University of Turin, and holds a Ph.D. in History of Philosophy at the Faculty of Philosophy of the University of Paris IV Sorbonne. His main fields of research concern Philosophy of Law, Legal Informatics and Information Ethics. Author of several books, he has widely published articles in Italian, English and French. He has recently edited, with Ugo Pagallo, the book: *Manuale di informatica giuridica e diritto delle nuove tecnologie,* Utet, Torino, 2012.

**Luciano Floridi** is Professor of Philosophy and Ethics of Information at the University of Oxford, Senior Research Fellow at the Oxford Internet Institute, and Fellow of St Cross College, Oxford. Among his recognitions, he was the UNESCO Chair in Information and Computer Ethics, Gauss Professor of the Academy of Sciences in Göttingen, and is recipient of the APA's Barwise Prize, the IACAP's Covey Award, and the INSEIT's Weizenbaum Award. He is an AISB and BCS Fellow, and Editor in Chief of Philosophy & Technology and of the Synthese Library. He was Chairman of EU Commission's "Onlife Initiative". His most recent books are: The Fourth Revolution—How the infosphere is reshaping human reality (OUP, 2014), The Ethics of Information (OUP, 2013), The Philosophy of Information (OUP, 2011), The Cambridge Handbook of Information and Computer Ethics (editor, CUP, 2010), and Information: A Very Short Introduction (OUP, 2010).

**Mireille Hildebrandt** holds the Chair of Smart Environments, Data Protection and the Rule of Law at Radboud University Nijmegen; she works as a legal philosopher at the Erasmus School of Law, Rotterdam and is part of the research group on Law, Science, Technology & Society studies at Vrije Universiteit Brussels. Hildebrandt co-edited e.g. Profiling the European Citizen (Springer 2008) together with Serge Gutwirth and Privacy, Due Process and the Computational Turn (Routledge 2013) together with Katja de Vries. Her research interests focus on the nexus of law, philosophy of law, and philosophy of technology.

**Hosuk Lee-Makiyama** is Director of the European Centre for International Political Economy (ECIPE). He has published several papers and articles for ECIPE since 2010 on international trade, multilateral governance, and the digital economy.

Prior to joining ECIPE, he was serving in the Ministry of Foreign Affairs of Sweden, serving as EU presidency chair in the World Trade Organization and the UN.

**Dawn C. Nunziato** is a Professor of Law at The George Washington University Law School and an internationally recognized expert in the area of free speech and the Internet. She recently published her book *Virtual Freedom: Net Neutrality and Free Speech in the Internet Age* (Stanford University Press).

Professor Nunziato has taught Internet law courses and lectured on Internet free speech issues around the world, including at Oxford University; the Munich Intellectual Property Law Center; Tsinghua University in Beijing; the Center for Studies on Freedom of Expression and Access to Information at the University of Palermo in Buenos Aires; and the Organization for Security and Co-Operation in Europe in Vienna. She has been an invited presenter and speaker at Yale, Harvard, Oxford, Georgetown, Vanderbilt, and University of Virginia, among other institutions.

Professor Nunziato served as Articles Development Editor of the Virginia Law Review and was the recipient of the Thomas Marshall Miller Prize, awarded to the outstanding member of the graduating class at University of Virginia Law School.

**Ugo Pagallo** is Professor of Jurisprudence at the Department of Law, University of Turin, since 2000, faculty at the Center for Transnational Legal Studies (CTLS) in London and faculty fellow at the NEXA Center for Internet and Society at the Politecnico of Turin. Member of the European RPAS Steering Group (2011–2012), and the Group of Experts for the Onlife Initiative set up by the European Commission (2012–2013), he is chief editor of the *Digitalica* series published by Giappichelli in Turin and co-editor of the AICOL series by Springer. Author of nine monographs and numerous essays in scholarly journals, his main interests are AI & law, network and legal theory, robotics, and information technology law (specially data protection law, copyright, and online security). He currently is member of the Ethical Committee of the CAPER project, supported by the European Commission through the Seventh Framework Programme for Research and Technological Development.

**Giovanni Sartor** is part-time full professor in legal informatics at the University of Bologna and part-time professor in Legal informatics and Legal Theory at the European University Institute of Florence. He obtained a PhD at the European University Institute (Florence), worked at the Court of Justice of the European Union (Luxembourg), was a researcher at the Italian National Council of Research (ITTIG, Florence), held the chair in Jurisprudence at Queen's University of Belfast, and was Marie-Curie professor at the European University of Florence. He has been President of the International Association for Artificial Intelligence and Law. He has published widely in legal philosophy, computational logic, legislation technique, and ICT law. He is co-director of the Artificial intelligence and law Journal and co-editor of the Ratio Juris Journal. His research interests include legal theory, logic, argumentation theory, modal and deontic logic, logic programming, multiagent systems, computer law, data protection, human rights.

**Mariarosaria Taddeo** is Fellow in Cyber Security and Ethics and the Department of Politics and International Studies at the University of Warwick and Research Associate at the Uehiro Centre for Practical Ethics, University of Oxford. Her recent work focuses mainly on the ethical analysis of cyber security practices and information conflicts. She has worked on issues concerning Philosophy of Information, Epistemology, Philosophy of AI and Applied Ethics. Her publications include several articles on online trust, cyber security and cyber warfare. She edited (with L. Floridi) a volume on The Ethics of Information Warfare (Springer, 2014). Dr. Taddeo is the 2010 recipient of the Simon Award for Outstanding Research in Computing and Philosophy and of the 2013 World Technology Award for Ethics. She serves in the executive editorial board of Philosophy & Technology and is the President of the International Association of Computing and Philosophy.

# Chapter 1
# The Right to be Forgotten: Dynamics of Privacy and Publicity

**Giovanni Sartor**

**Abstract** The passage of time may affect the balance of the interests involved in the processing of personal data. In particular it may have an impact on the trade-off between publicity interests and privacy interests with regard to information made available online. Changes in this trade-off may justify a transition in the legal status of the same piece of information: what was legitimately distributed at an earlier time may no longer be legitimately provided to the public at a later time. This idea is at the core of the so-called "right to be forgotten", namely, the data subject's right to obtain, at a later time, the erasure of personal information that originally was legitimately processed. This right has been endorsed in a number of judicial decisions in various EU member states, and has been explicitly affirmed in the Proposal for a General Data Protection Regulation, presented by the EU commission in 2012.

Here I propose a method for modelling the evolution of the privacy and publicity interests through time, and for assessing the impacts of a discipline of the right to be forgotten on the online distribution of information. I will distinguish different trends in the trade-offs between privacy and publicity, and more generally between the interests that would be promoted by a certain processing and those that would be demoted by it. I will argue that in cases where there is a reversal-time, mainly a time when the first interests, originally prevailing, are outweighed by the latter, the law may direct controllers (or processors) to stop or change the processing around that time, and I will consider ways of achieving this outcome.

G. Sartor (✉)
Legal Informatics, University of Bologna, Bologna, Italy
e-mail: Giovanni.sartor@gmail.com

Legal informatics and Legal Theory,
European University Institute of Florence, Florence, Italy

L. Floridi (ed.), *Protection of Information and the Right to Privacy – A New Equilibrium?*,
Law, Governance and Technology Series 17, DOI 10.1007/978-3-319-05720-0_1,

## 1.1 Data Protection and the Passage of Time

The right to be forgotten concerns the way in which the passage of time affects the legitimacy of data processing: processing operations that were legitimate up to a certain point in time may become illegitimate after that time.[1] This may happen in particular when a processing is based on legitimate interests (of the controller or of third parties) or on social values, but such interests and values, while originally prevailing over privacy-related interests of the data subject, are at a later time outweighed by the latter interests. In particular, publicity interests broadly understood (interests pertaining to freedom of expression and right to information, as well as to values such as democracy, transparency, informed public deliberation, etc.), while prevailing at the time when a piece of information was published online, may be outweighed at a later time by privacy interests broadly understood (data protection, reputation, identity, dignity, the right to a fresh start, etc.) of the concerned data subject. Consider, for instance, journal articles, blog posts, photos or videos published online, reporting events the involvement in which negatively affects the data subject (for instance, crimes, failed business activities, scandals, etc.). While individual freedoms as well as the social value of information and public debate justify the distribution of such information (even when the data subject would strongly prefer that it were not made accessible to the public) at the time of the event, it may happen that at a later time the balance of the interests changes, so that the information should no longer be distributed.[2] In the following a detailed analysis of the different ways in which this balance is affected is proposed.

### *1.1.1 Achievement of the Purpose*

Let us first address the case when at a certain point the purpose of processing has been fully achieved (for instance when a commercial transaction has been successfully completed). Let us assume that at this point there is no longer any relevant justified interest in continuing processing (no interest pertaining to the completed transaction). The pattern characterising such a situation is presented in Fig. 1.1. The horizontal axis represents the passage of time, from the initial moment when the processing started (time 0). The vertical axis represents the legal significance of the impacts on the interests at stake, on the one hand the importance of the benefits with regard to the interests being advanced by the processing (pertaining, for instance to security, or public health, or to the implementation of a contract) and on the other hand the importance of the loss in privacy.

[1] On digital forgetting, see Mayer-Schoenberger (2009). On the right to be forgotten, see among the others, Werro (2009); Weber (2011); Koops (2011); Rosen (2012); Ambrose and Ausloos (2013); Erdos (2013). For an introduction to the General Data Protection Regulation, see for all Kuner (2012). On the philosophical aspects of information, identiy and privacy, see Floridi (2006).

[2] On balancing in the law, also for references to the huge legal literature on the matter, see Sartor (2013).

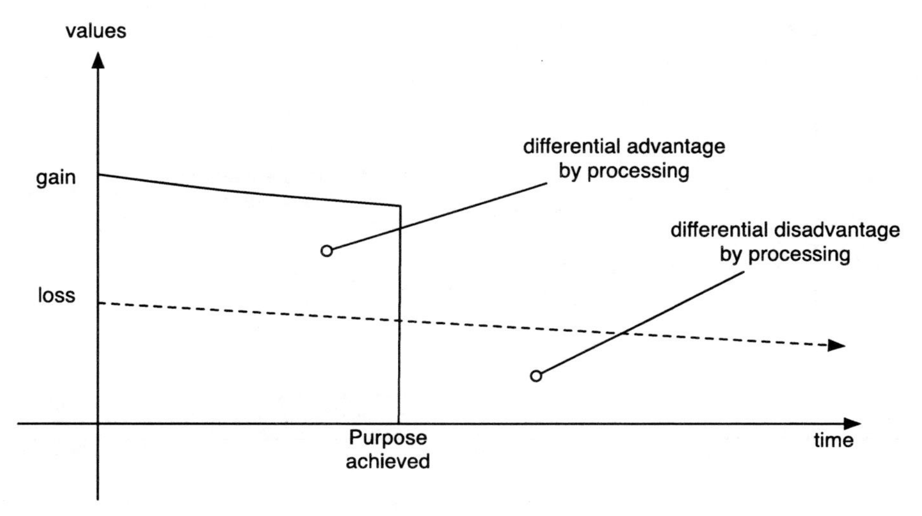


**Fig. 1.1** Achievement of processing's purpose

This representation is based on the assumption that each impact (gain and loss) with regard to the achievement of a value has a certain positive or negative legal force (significance or importance for the law), and that such forces can be added so that their aggregate importance can be compared. The aggregate force of the differential benefits provided by a processing (with regard to the values it is contributing to) is represented by their height of the curve named "gain from processing", while the aggregate force of the differential drawbacks provided by the same processing (with regard to the values it is detracting from) is represented by the height of the curve names "loss from processing". The assumption that aggregate pull of such impacts can be compared follows from the very idea of a proportionality assessment, namely, the assessment of the merit of choice by comparing its contribution to legitimate goals (interests, values) against its interference on other rights and values, though proportionality involves a further aspect, besides balancing, namely the idea of necessity, i.e., the non-availability of a less infringing way to equally achieve the legitimate goals.

Note that even when all conditions for legitimate processing are satisfied, a loss in privacy still takes place, which is, however outweighed by the benefits provided by the processing. After the achievement of the main goal this is no longer the case.

More precisely, the trade-off involved in the processing, namely the net outcome we obtain by subtracting the loss in privacy from the benefit to other interests is positive up to the time when the goal is achieved, then it becomes negative since the loss in privacy is no longer fully compensated by the achievement of other (legitimate) goals.

After that point there is no differential merit in continuing the processing.

A different context is shown in Fig. 1.2, where we see that after the main purpose of processing is obtained there may still be minor purposes to be achieved, by

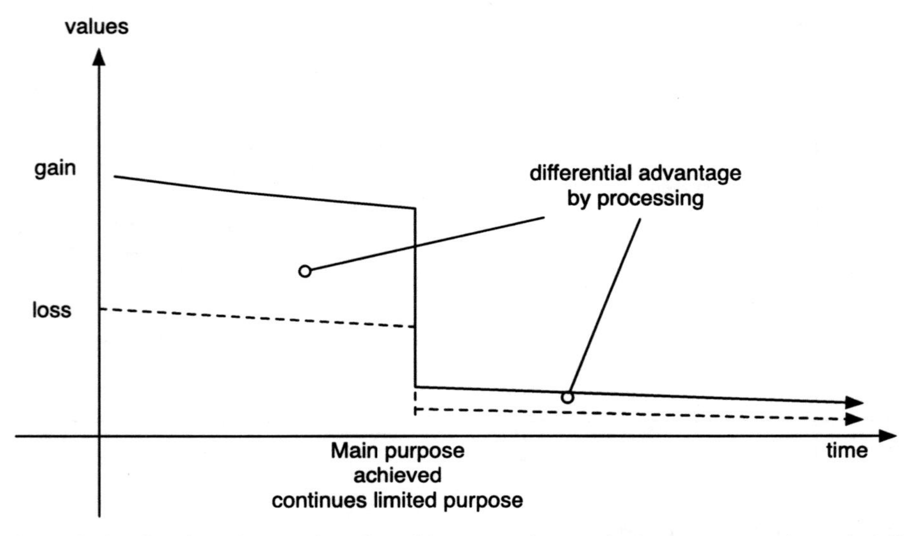


**Fig. 1.2** Continuation of processing after achievement of processing's purpose, negative trade-off

processing the data in ways that reduce the interference with privacy. Consider for instance the case when data about a client may need to be kept separately as a proof of a transaction, to be accessed only in case an issue emerges. Assume that this particular kind of processing still involves a small privacy loss, but this loss is inferior to the advantage brought about by the limited processing, after the end of the primary usage of the data. Then, according to the balance of interests, the limited processing may justifiably continue after the complete achievement of the main goal.

Note that it is not always the case that the minor advantage obtained by a further limited processing justifies a reduced interference with privacy. The minor advantage provided by the limited usage may be outweighed by the remaining loss to privacy, as indicated in Fig. 1.3. In such a case, the processing should be completely terminated as soon as the purpose is achieved. Thus for instance, it may be argued that while maintaining data from street cameras in encrypted form (possibly with additional security measures to protect secrecy), would involve a lesser impingement on privacy than keeping the data unencrypted, this lesser impingement would still outweigh the benefit to security. Therefore such data should rather be erased after a short retention time.

### *1.1.2 Decreasing Impacts, Persisting Priorities*

In the cases we have just considered, we assumed that at a certain point in time the purpose of processing is achieved, which involves the termination or the sudden reduction of the benefits associated with processing. When the interference of privacy results from the distribution of personal information for purposes pertaining

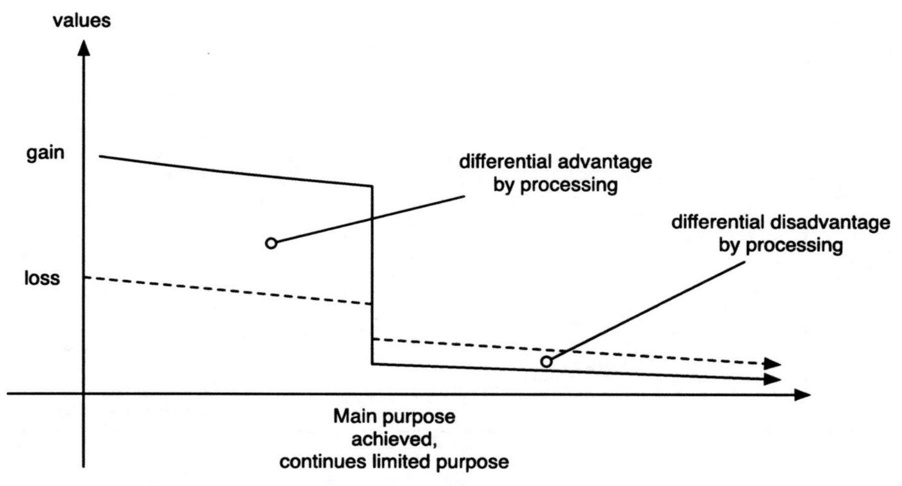


**Fig. 1.3** Continuation of processing after achievement of processing's purpose, positive trade-off

to journalism, or more generally to freedom of expression, the trend is likely to be different: rather than a sudden downward jump, there is a continuous diminution.

In such cases as time goes by, the processing usually has a continuously decreasing impact on both publicity-interests and privacy-interests. In fact, older personal facts are usually less significant for both the public and the data subject. In particular, older information about a person gives a less significant clue on what a person is now, and therefore is less relevant both for those who want to know about that persons and for the person herself.

This assumption, however, is compatible with various arrangements of the interests at stake.

First of all, the loss concerning privacy-interests may override the gain concerning publicity-interests (the benefit resulting from the distribution of the information) from the very beginning. Consider for instance the publication online (on a blog or a journal) of data concerning health or sexual preferences of a person, when this information has no or little significance for her social role, while having some significance for people's curiosity. In such cases, as shown in Fig. 1.4, the loss to privacy is at no time compensated by gain in freedom of expression/information. Therefore the processing should be always forbidden.

The same considerations also apply to cases where security interests are overridden from the beginning by privacy interests; consider for instance the issues involved in putting cameras in changing rooms, or even in meeting rooms where no specific security dangers are present.

The situation is completely overturned in the scenario represented below in Fig. 1.5, where instead interests in publicity (that a certain piece of information is distributed and accessed) always outweigh the privacy interest of the data-subject. As an example, consider personal information concerning a person having an important public role, information that is relevant to that role and maintains a permanent

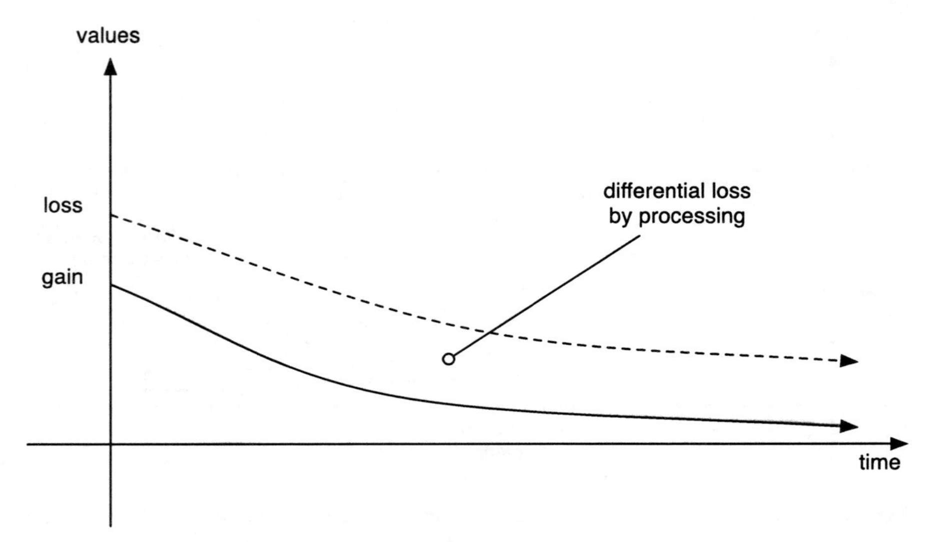


Fig. 1.4 Processing with persistent negative trade-off

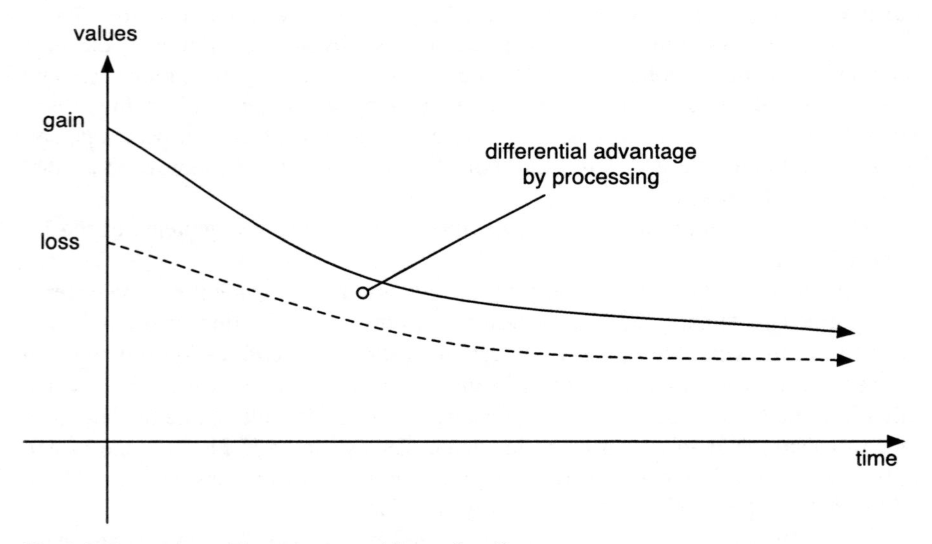


Fig. 1.5 Processing with persistent positive trade-off

exemplary or historical significance. Also in such a case the time-induced reduction in the impacts on both publicity and privacy interests does not change the relative position of the two: the first always outweighs the latter. Thus processing (and in particular distribution) should always be allowed regardless of the passage of time.

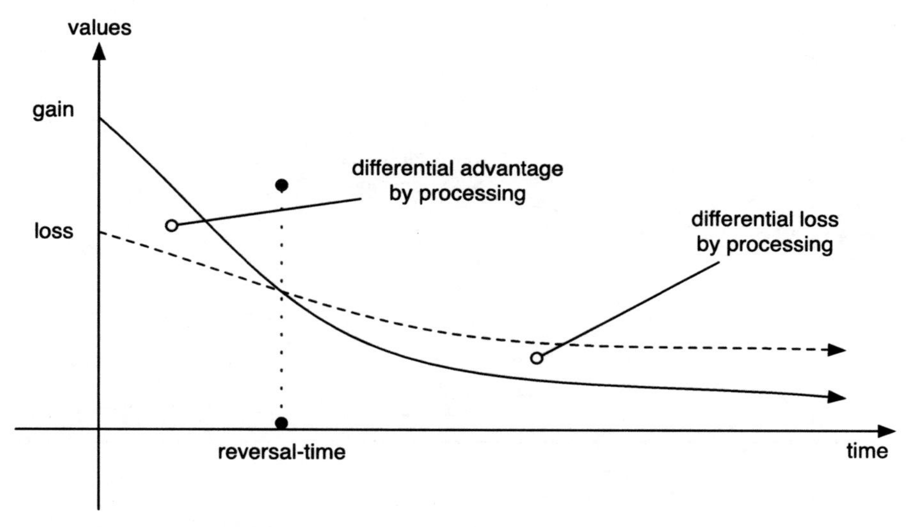


**Fig. 1.6** Reversal of the trade-off

## *1.1.3 Decreasing Importance of the Impacts, Changing Priorities*

Let us now address the typical context for the right to be forgotten or to oblivion: up to a certain point in time publicity prevail, and after that point privacy takes the lead, as shown in Fig. 1.6. In addition to the assumption above, i.e., that both privacy and publicity interests have a downward trend, we need to make a further assumption to capture these cases. This is that idea that as time goes by, the decrease rate of the impact on privacy-interests is generally smaller than the decrease rate of the impact on publicity interests. This can be explained by the fact that while the public significance of information is related to its actuality, old information about a person tends to remain significant for that person, since it continues to have an effect on how that person is publicly perceived

In such cases, both impacts on freedom of expression and on privacy decrease as time goes by, but the diminution of the impact on freedom of expression proceeds at a steeper pace, so that while at the beginning the benefit outweighs the loss, at a certain point in time there is a change: the loss in privacy outweighs the benefit in freedom of expression/security. This is the point where, arguably, the data should be forgotten: the maximization of the overall differential outcome is obtained by switching at that time from distributing to erasing the data. In this way we keep differential advantage in favour of publicity obtainable before the reversal-time while avoiding the differential loss on privacy of privacy that would take place after that time.

A similar situation occurs where certain information relevant to security (e.g. information taken from cameras on shops and streets) loses most of its significance after a short time (since arguably the effects of crimes usually can be immediately

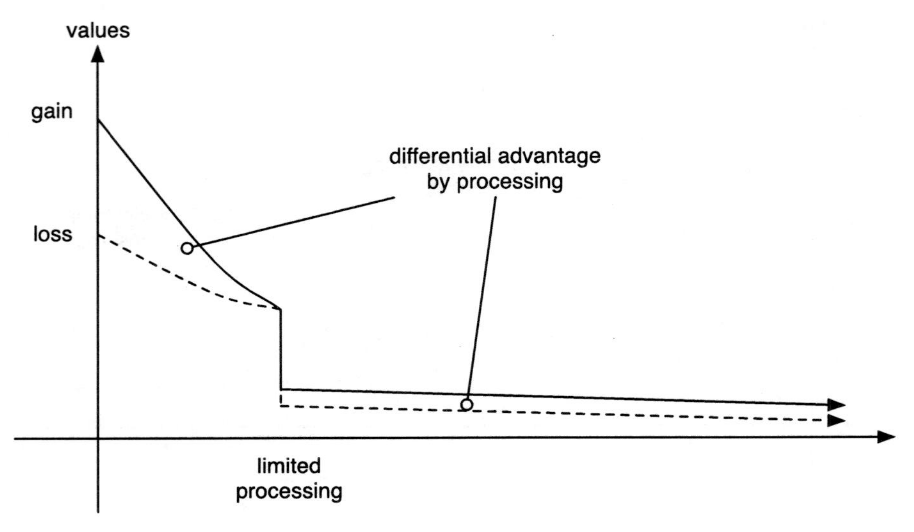


**Fig. 1.7** Limited processing from the reversal-time

detected, so that investigation can be performed a short time after the crime was committed) while continuing to have a negative impact on the privacy of the persons whose images are stored.

A different solution may however be possible in certain cases, namely a change at the reversal-time in the ways in which the data are processed. With regard to information distributed on line, this may be obtained by ensuring that the stored data is less easily usable and accessible, leading to the pattern shown in Fig. 1.7. For instance, a number of judicial decisions have recognized that newspapers are not required to delete articles containing old personal information. It may be sufficient that they store these articles in separate archives, so that they are no longer accessible through general searches on the web, but they are obtainable only by specific queries addressed to such archives. When privacy is opposed to freedom of information in such a constellation of interests, we may possibly speak of a right to (partial) concealment or to minor accessibility, rather than a right to erasure. In other cases it may be preferable to complete the information, rather than deleting or concealing it. This is the case in particular when the information may be misleading, unless accompanied by further data that changes its significance (e.g. information about prosecution without the indication that the accused person was later acquitted).

## 1.2 The Legal Regulation of Time-Dependant Priorities and the Right to Oblivion

Various legal arrangements can be adopted to deal with the connection between legitimate processing and the passage of time.

Firstly, the principles of purposiveness and necessity of personal data require that the data be erased when all purposes are achieved for which it was processed.

Secondly, when the data subject's consent provides the only legitimate ground for processing, then by making this consent revocable the law gives the data subject the power to make the processing unlawful (since it would lack the needed legal basis) at any time. Thus the data subject's assessment that she prefers to forbid processing (rather than allowing its continuation) provides, in such cases, a sufficient criterion for the impermissibility of processing her data. Interestingly this empowerment of the data subjects may also a disempowerment, namely, it entails her inability to unconditionally transfer her personal data to another, and thus her inability to obtain the price or benefit that may result from such an unconditional transfer.

Thirdly, when the processing is justified by a legitimate interest of the controller or a third party, then the law may provide for maximum retention times through appropriate rules, meant establish the borderline between lawful and unlawful processing, according to a generalised appreciation of how the balance of interests is likely to evolve over time. This kind of regulation may be most appropriate to address those contexts where after a defined time span it is unlikely that the processing's positive impacts still outweigh its negative impacts on privacy. Consider for instance, the storage of footage from street cameras, the storage of data concerning innocent people involved in police investigations, the retention of data concerning unsuccessful job applications, the registration of data on phone or e-mail communications, etc.

In the absence of rules establishing express deadlines, the law may confer to an authority (e.g. a court or a data protection authority), the task to establish in whether privacy interests have outweighed the competing interests in individual cases. In the context of such judgments, the position of the data subjects can be strengthened by burdens of proof or of argumentation, so that the processor has the burden of showing that the interests supporting the processing still outweigh privacy interests, to be able to continue processing and avoid sanctions.

With regard to the object of our inquiry, i.e., the distribution on line of content relevant to the public, it seems to me that the first criterion for erasure (full achievement of purpose) cannot apply, since the purposes of the processing (informing and being informed) are still present at any later time: the processing party still has an interest in distributing that content and some members of the public may still be interested in accessing it (though such interests are likely to be diminished to some extent).

Similarly, also the second criterion for erasure does not apply, since the distribution of the content is not justified only by the data subject's consent. Freedom of expression and freedom of information provide independent grounds for distribution.

Thus we need to focus on the third case, namely, the supervened unbalance between publicity and privacy interests, the first having been outweighed by the latter. In particular we need to examine:

- whether the prohibition to distribute the data or at least its enforceability should start when the data subject requests removal or when there is a removal decision by a competent authority; or

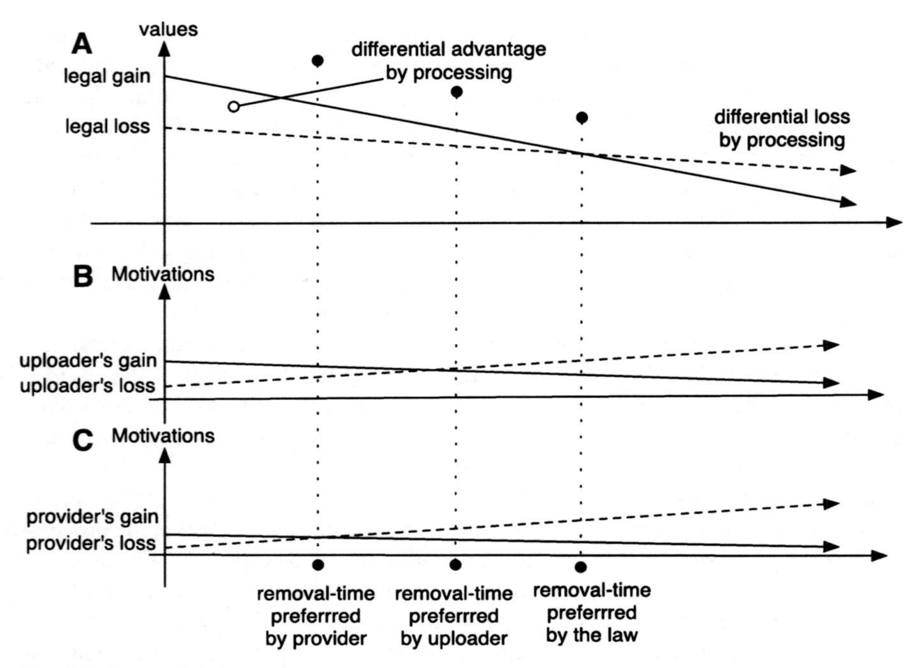


**Fig. 1.8** Removal times

- how the law should intervene to enforce such a prohibition, through what sanctions or injunctions.

By examining these aspects we will be able to understand how a legal regulation should influence the behaviour of the parties and identify critical aspects. I shall argue that exactly in this regard, the discipline provided in the proposed Data Protection Regulation appears to be inadequate.

### *1.2.1 The Interests at Stake in the Flow of Time and Possible Sanctions*

Let us return to the situation we analysed in Fig. 1.6 above, namely the case when the passage of time brings about a supervened prevalence of privacy over publicity.

In Fig. 1.8 a linear relationship is assumed between the represented interests and time, for simplicity's sake.

In the part A of Fig. 1.8, the privacy line starts at the higher level, but decreases more rapidly than the publicity line, so that at a switch point the two lines cross: from that point on, the damage to privacy is no longer compensated by the benefit to publicity interests.

Figure 1.8 also contains a representation of the motivation of the uploader (part B) and of the host provider (part C), both of which are also decreasing, but remain positive (assuming that no sanctions are provided).

The uploader's motivation includes the economic gains the uploader expects from distributing the information (as is the case for newspapers and websites getting subscriptions or advertising), but also includes the moral and social importance the uploader attributes to providing such information. Abstracting from different individual attitudes, we may assume that the motivation for distributing information is measured by the maximal personal loss one would be ready to sustain for not distributing it, regardless of the grounds that explain this attitude. Consider, for instance, the situation of a person who has to decide whether to upload on a blog information concerning a political or economic scandal. This person knows that this may cause him some personal advantage (reputation, some chances of having a political role in the future, possible some financial gain resulting from the fact of attracting people to the blog) but larger personal losses (e.g., losing possible contracts, missing career advancements, even putting at risk one's life or freedom, etc.), and he also knows that this information would be highly beneficial to the public, contributing to curb the plight of corruption, while damaging the data subject. The motivation of such a person would likely be measured neither by the mere trade-off of personal gains and losses, nor by adding to this trade-off the full amount of the expected (net) public benefit. It would rather be measured by adding to the trade-off of personal gains and losses a quantity expressing how much he personally values the moral/social merit of his action, i.e., what additional personal losses he would be ready to sustain to accomplish this action.

The motivation of providers is similar to that of uploaders, while being generally lower. Providers host huge amounts of document, so that while they are interested in having a discipline that allows them to keep their information on line, or that at least does not require them to control it, they have a small interest in maintaining a single piece of information.

It may seem that to induce the processing parties (uploaders and host providers) to remove the information as soon as the switch time is reached, distribution after the switch time should be sanctioned. The sanction, to be effectively deterrent, should be as high as to prompt most such parties to behave as requested. In general, we may assume that the sanction may include the compensation of the damage to the data subject, as requested by the Art. 23 of the EU data protection directive—a compensation which according to certain national regulations, such as the Italian one, also includes non-economic damage—plus additional administrative sanctions, as established by national legislation, and as required by the Proposal of a general data protection regulation at art.

If such a sanction were always to be imposed upon a processing only after the point in time where the balance between privacy and publicity interests is reversed, and the processing party knew exactly where this point is located, such a discipline would induce the behaviour that maximises the achievement of legal values. Before the reversal-time uploaders and providers would leave the material on line, since they could enjoy the benefits resulting from the distribution of the information without encountering any legal sanction; after that point, they would take it down, since continuing to distribute the information would expose them to the obligation to compensate damages of the data subject, and to any further sanction established by data protection law (or other legal rules).

This analysis however, is fallacious, since it fails to consider a further aspect. The processing parties may be uncertain as to whether distributing certain information at a certain point in time provides a positive or rather a negative balance between publicity and privacy interests, being therefore lawful or rather unlawful. Or in any case they may be uncertain as to how the competent decision maker will judge the issue.

This means that even before the switch time, the expected sanction will not be 0. Rather, the expected sanction for keeping the information on line at a time $t$ will result from multiplying the amount of the sanction at $t$ for the probability of being sanctioned, which corresponds to the probability that the decision maker will ex post consider that at time $t$ the con-processing interest already outweighed the con-processing interests. Such a probability is increasing as time goes by, if we assume that it is inversely correlated to the difference between impacts on publicity interests and on privacy interests: as this difference diminishes (and then is reversed), it becomes more and more probably that the adjudicator may consider that privacy interest outweigh publicity interests. Thus, assuming that the amount of the sanction does not increase over time, we have that the expected sanction for the continued distribution increases over time, as is represented by the upward broken line in the lower part of Fig. 1.8.

According to Fig. 1.8, the processing parties will be motivated to take down the information before the time when this is legally preferable. Let us now consider in detail why this seems plausible, at least when the loss provided by the sanction is considerably superior to processing party's motivation to keep the information on line.

For this purpose we have to analyse in detail the judgement through which a processing party tries to assess whether the information, which was previously assumed to be lawful, on the basis of a positive trade-off at time $t_0$, may have become unlawful at a later time $t_1$. This judgment can only be a probabilistic one, since the processing party will be able to make a very approximate estimation of both the gain in publicity interests and the loss in privacy interest taking place at time $t_1$. This uncertainty is likely to increase the more the time line is approaching the reversal-time.

To identify the point in time from which the party will no longer be motivated to continue to distribute the information, we need to consider three aspects: the loss that the party would suffer in case the data were considered to be illegal (publicity interests being outweighed by privacy interests), the probability that the party assigns to the data being considered illegal, and the motivation that the party has for leaving the material on line.

The first element is the total loss expected in case the data were considered to be illegal at the considered time $t$: this is the amount given by the compensation of the privacy-damages of the data subject, plus possible fines.

The second element is the subjective probability (by the concerned party) that indeed the information will be considered to be illegal (according to the judgement of the competent authority, based on the evaluation of the balance of interests). Let us consider the case of personal information distributed to the public, and relevant to the public to such an extent as to make its distribution originally permissible. At the reversal time and around it there will likely be a high uncertainty on whether the reversal time has been reached, or will be considered so by the decision maker. Now, by combining the sanction and its probability, we obtain the expected loss of the processing parties, the uploader and the provider. More exactly, the expected loss of such a party resulting from the distribution of information possibly considered to be illegal is obtained by multiplying the total loss of that party, in case the data were considered illegal, for the subjective probability that indeed the data could be considered illegal (which is likely to be 50% around the reversal-time).

The third element is the motivation that the processing parties have for leaving the material in line, measured by what threatened financial loss (or promises reward) would be sufficient to induce such parties to remove the information. This motivation has to be compared with the amount of the expected sanction for illegal distribution, which will include economic damage to the data subject, plus possibly the moral damage to the latter and further administrative or criminal sanctions. It is likely that the motivation of both the data subject and the provider will be much lower than this sanction, in the case of distribution of information to the public. In fact the economic interest of the uploader in having a particular content distributed is likely to be small, especially when the provider is not a professional journalist. Similarly, also the economic interest of the provider in publishing a single piece of information is likely to me very small, being limited to the marginal income stream related to hosting that particular piece of information, among the vast amount of information which is available on the provider's platform. It is true that the motivation of an uploader may also include social-political goals related to distributing to the public critical information, but the importance of such information for the public is likely to be internalised only to a limited extent by the uploader himself. In fact, while the impact on the interests negatively affected by the on-line distribution of information will be reflected in the economic and moral damage suffered by the data subject, the interests that are benefitted by the distribution of the information will not be internalised to the same extent by the uploader and the provider.

In general, given the analysis just provided of these three elements (total loss, subjective probability of it, and motivation), we can conclude not only that there is a point in time when the expected sanction will outweigh a party's motivation to distribute the material, but also that—with regard to the continued on-line distribution to the public or originally permissible information—this point in time with generally be earlier than the reversal-time. Thus a legal provision sanctioning continued publication after the reversal time will induce the processing parties to withdraw the material prematurely, i.e., at times when publicity interests still outweigh privacy interests. This anticipation will be larger when the motivation is smaller, the sanction is larger, or the uncertainty is greater. Note that to have this effect, the sanction does not need to be extremely severe or fully certain: it suffices that the sanction, discounted by the probability of not being punished, overrides the motivations of the parties. Also a punishment limited to damages (in particular when also moral damages are included) may have such a result. However the anticipatory effect of potential sanctions is greatly enhanced when the for the sake of general deterrence, the sanction becomes very large, to an extent exceeding damage. Thus, the provision of Art. 80 in the Draft Regulation, according to which a supervisory authority "shall impose a fine up to 500,000 €, or in case of an enterprise up to 1 % of its annual worldwide turnover" for violation of the right to be forgotten, if applied also to the on-line distribution of information to the public, may induce the removal of information that should continue to be distributed for the sake of freedom of speech and information.

Premature self-censorship is not uniquely dependant on the assumption that the concerned individuals are deciding rationally on the basis of their expected utilities: it would likely happen even to an even larger extent (at least as the reversal-time approaches) if the concerned individuals were assumed to decide on the basis of the most popular behavioural approach, i.e., prospect theory, since the latter emphasises the significance of sufficiently probable potential losses.

## 1.3 Conclusion

I have considered how the passage of time may affect the interests that are involved in the processing of personal information. After distinguishing various possible trends I have focussed on the continued on-line distribution of information whose publication was originally permissible.

I have argued that sanctioning the distribution from the point in time when the privacy-related interests outweigh the publicity-related ones is likely to lead to anticipated removal. To avoid this result, different approaches are available.

The first approach consists in establishing the obligation to remove the information after a definite maximal retention time, an obligation which may be supported by strong sanctions without inducing premature removal, since the processing party should have no doubt as to when the deadline falls. However such an approach does not seem to be appropriate with regard to the online publication of information, since the time span in which the balance of interests favours publication is very variable depending on the kind of information at issue and the social context in which it is distributed.

The second approach consists in refraining from punishing the persistent publication of originally permissible information, even after a removal request, while allowing the remedy of a removal injunction from a competent authority. This may err on the side of non-publication, involving that there is no incentive for the processing party to remove the information before such an injunction.

A third approach consists in sanctioning the continued distribution of information only when it appears to be in bad faith, namely, incompatible with a reasonable belief that the continued publication is permissible. Such an approach could be refined by differentiating the condition for an injunction and for a sanction (compensation and fine). The injunction should be conditioned to the mere reversal of the balance of legal interests, while the sanction should require the unreasonableness of the assumption that the reversal has not yet taken pace, or the moral certainty that a reasonable decision-maker would assume that the reversal has taken place.

With regard to the regulation of the right to be forgotten, I would argue, a combination of the latter approaches is needed, possibly with a preference for the third approach with regard to the uploader and for the second approach with regard to the provider.

## References

Ambrose, M. L., and J. Ausloos. 2013. The right to be forgotten across the pond. *TRPC, Journal of Information Policy* 3:1–23.

Erdos, D. 2013. Freedom of expression turned on its head? Academic social research and journalism in the European privacy framework. *Public Law* 1:52–63.

Floridi, L. 2006. The ontological interpretation of informational privacy. *Ethics and Information Technology* 7:185–120.

Koops, B.-J. 2011. Forgetting footprints, shunning shadows. A critical analysis of the "right to be forgotten" in big data practice. *SCRIPTed* 8:1–28.

Kuner, C. 2012. The European commission's proposed data protection regulation: A Copernican revolution in European data protection law. *Privacy and Security Law Report* 11:1–15.

Mayer-Schoenberger, V. 2009. *Delete: The virtue of forgetting in the digital age*. Princeton: Princeton University Press.

Rosen, J. 2012. The right to be forgotten. *Stanford Law Review Online* 64.

Sartor, G. 2013. The logic of proportionality: Reasoning with non-numerical magnitudes. *German Law Journal* 14:1419–1457.

Weber, R. H. 2011. The right to be forgotten: More than a Pandora's box? *Journal of Intellectual Property, Information Technology and E-Commerce* 2:120–130.

Werro, F. 2009. The right to inform v. the right to be forgotten: A transatlantic clash. In *Liability in the third millennium*, eds. A. Colombi Ciacchi, C. Godt, P. Rott, and L. J. Smith, 285–300. Baden-Baden: Nomos.

# Chapter 2
# Legal Memories and the Right to Be Forgotten

**Ugo Pagallo and Massimo Durante**

**Abstract** The paper examines the current debate on the right to be forgotten in connection with three different issues that revolve around: (i) the construction of individual identities; (ii) how individual and collective memories are intertwined; and, (iii) different forms of oblivion vis-à-vis the idea of forgiveness. The aim is to offer a normative stance in terms of "fair memory" and "difficult forgiveness." From a philosophical viewpoint, attention is drawn to the dual status of the past, i.e., that which is not any longer and what Paul Ricoeur used to call the "existent state." From a legal perspective, focus is on how to strike a balance between the subjective claim to be forgotten and further rights of the legal system. From a political outlook, what is at stake concerns the mediation between the relational structure of the law and the inter-subjective nature of forgiveness. Today's debate has to match up with all these aspects of the right to be forgotten.

## 2.1 Introduction

The "right to be forgotten" should not be considered as a new right in the legal domain and still, the information revolution has forced both legislators and courts to rethink the ways in which such right must be understood. On the one hand, the right to be forgotten can properly be traced back to the traditional right to respect for private life, e.g., Article 8 of the European Convention on Human Rights ("ECHR"), much as the protection of personal data. Consider, for instance, Article 6 (e) of the European directive on data protection, i.e., D-95/46/EC and the provisions on data that should be kept for no longer than is necessary for the purposes for which such data was collected. On the other hand, the information revolution has profoundly impacted on this framework because of the persistence, replicability, scalability and searchability of information on the internet, along with the de-contextuability and

U. Pagallo (✉) · M. Durante
Law School, University of Turin, Torino, Italy
e-mail: ugo.pagallo@unito.it

M. Durante
e-mail: massimo.durante@unito.it

L. Floridi (ed.), *Protection of Information and the Right to Privacy – A New Equilibrium?*, Law, Governance and Technology Series 17, DOI 10.1007/978-3-319-05720-0_2,

re-combinability of content of individual messages. Accordingly, on 25 January 2012, the European Commission has presented the proposal for a new regulation on data protection (2012/0011 (COD)), whose Article 17 on the "right to be forgotten and to erasure" has sparked much controversy. Pursuant to Article 17(1) of the proposal, "the data subject shall have the right to obtain from the controller the erasure of personal data," in all the cases in which the data are no longer necessary, the data subjects withdraw consent, or object to the processing, and so forth. In addition, the proposal aims to introduce new obligations for data controllers: whereas the latter shall carry out the erasure without delay, and restrict processing of personal data where their accuracy is contested by the data subject, the data controllers "shall take all reasonable steps, including technical measures, ... to inform third parties which are processing such data, that a data subject requests them to erase any links to, or copy or replication of that personal data" (Art. 17.2 of the Proposal).

Meanwhile, courts and tribunals have been very active: suffice it to recall the decisions of both the Tribunal de grande instance de Paris and the Italian Corte di Cassazione, vis-à-vis the doubts of the Audiencia Nacional in Madrid. First, on 15 February 2012, an *ordonnance de référé* of the Tribunal de grande instance de Paris ordered the search engines Google.com and Google.fr to remove from their services all of the links that could trace the plaintiff Diana Z. back to her previous life (and artistic work) of porno actress. Although this decision looks highly problematic in light of the clauses of immunity for internet service providers ("ISPs"), such as Article 15 of D-2000/31/EC, namely the European directive on e-commerce, this trend was confirmed two months later in Italy. On 5 April 2012, the third section of the Court of Cassation in Rome established in case n. 5525, that online news archives, such as those of the Italian newspaper *Il Corriere della Sera*, should be kept updated, in order to enforce the individual right to be forgotten: recall Article 6 (d) D-46/95/EC and the provision that personal data should be "accurate and, where necessary, kept up to date." On 9 March 2012, similar problems were already discussed before the Audiencia Nacional in Madrid (C-131/12): the issue concerned once again the obligations of search engines as providers of content in relation to Directive 95/46/EC on data protection. Significantly, the approach of the Court was more problematic and, as a result, a Reference for a preliminary ruling from the Audiencia Nacional was lodged before the EU Court of Justice in Luxembourg. The doubts of the Spanish Court can be summarized with two main points: according to no. 2.3 of the reference, can the Spanish Data Protection Agency "impose on the search engine of the Google undertaking a requirement that it withdraw from its indexes an item of information published by third parties, without addressing itself in advance or simultaneously to the owner of the web page on which that information is located?" Moreover, in the phrasing of no. 3.1 of the reference, do "the rights to erasure and blocking of data, provided for in Article 12(b), and the right to object, provided for by Article 14(a), of Directive 95/46/EC, extend to enabling the data subject to address himself to search engines in order to prevent indexing of the information relating to him personally"?

Leaving aside old and new obligations of internet service providers (Pagallo 2011; Reed 2012; etc.), focus is here on the philosophical reasons of today's legal debate, namely the role of oblivion in social interaction and the increasing amount of data and information, available on the internet, that shape individual identity. By reversing Jonathan K. Foster's thesis that "we are what we remember" (Foster 2008), it seems fair to affirm that oblivion plays a crucial role in people's lives, since it allows bearing the weight of the past and reprogramming the future. Whereas memory allows "to speak, read, recognize objects, orient ourselves in our environment or maintain contact" (Foster 2008), the right to be forgotten may be understood as a way to protect individual autonomy and permit people to remove their data, that is, "the traces of the past" that regard or affect the individuals. As it occurs with other provisions of the data protection framework, such as the transparency and fairness of data processing, whether data may be used and processed, etc., no violation of personal identity is thus necessary to grant the power to remove, or see removed, the data. Rather, such right to remove, or see removed, the data would be an expression of the autonomy with which every individual should be able to present and describe herself.

On this basis, we may say that both memory and oblivion cooperate to the construction of the personal identity as two sides of the same coin. In legal terms, this means that the protection of the right to speech, much as freedom of information and access, of press, etc., should go hand in hand with the protection of the right to be forgotten. But, how far should this latter power go? How about possible limitations of the right to be forgotten in the name of "legal memories"? How could a balance between memory and oblivion be properly struck?

In order to offer a hopefully comprehensive view on these issues, the paper is presented in four parts. Next, focus is on the philosophical status of the past: although it seems obvious that both memory and the right to be forgotten have to do with the dimension of the past, the meaning of such past is not evident. The confusion that often affects the current debate undermines the opinion that legal systems shall recognize and protect an individual right to be forgotten. This latter argument is deepened in connection with three different topics: the construction of personal identity in Sect. 3, the relation between individual and collective memories in Sect. 4, and the different forms of oblivion vis-à-vis the politics of forgiveness in Sect. 5. In each section, we examine what advocates and critics of the right to be forgotten argue, so as to offer a normative standpoint in the conclusions of the paper. Of course, our analysis does not aim to consider all of today's scholarly research but rather, we pay particular attention to the work of Paul Ricoeur. The French philosopher brilliantly wrote on memory and oblivion throughout his work, showing particular attention to the convergence between theory and practice with specific reference to the field of the law. In light of Ricoeur's meditation, the purpose is to grasp why everything in life is memory, apart from the thin line of the present and eventually, some aspects of the right to be forgotten.

## 2.2 The Status of the Past

In *La marque du passé* (1998), Ricoeur examines the nature of the past and its philosophical status, in order to prevent a usual mistake, namely to conceive the past as an entity, or as a place in which the experiences are deposited once they go by. Contemplate the Greek metaphor of the impression that a seal leaves on the wax: the metaphor refers to the recording and preservation of the traces of the past, which are necessary for the work of memory. Starting from such traces, the past could be reconstructed in the evocation of what is gone. However, memories do not remain unchanged over time: on the contrary, the act of recalling should properly be grasped as an active force (Foster 2008). After all, traces are not signs (Durante 2011): the thought of the past requires that the remembering subject makes sense of such traces via an active reconstruction. This latter stance warns against the risk of reifying the language—and this also may be the case of the legal language—that quite reductively assumes the concepts of memory and oblivion in terms of inscription, retention and deletion of data, as the traces of the past. By giving prominence to the materiality and availability of the data, rather than the more complex process of giving them meaning, this approach fails to grasp what is philosophically at stake with the nature of the past, i.e., what Ricoeur used to describe as "the enigma of the past," a puzzle between "what is no longer" and "the existent state" (Ricoeur 1998).

The nature of the past is indeed twofold, because it should be grasped in both a negative and positive ways. Whereas the negative conception of the past refers to that, which is lost forever, irretrievably deleted by the action of time, the positive concept sheds light on the past as something that remains, since we cannot pretend that nothing happened. This tension, according to Ricoeur, has been adequately examined by Martin Heidegger's meditation on time and the "analytic of existence," namely the analysis of what constitutes or defines our identity. By distinguishing between the past as *Gewesenheit*, that which was, and *Vergangenheit*, that which is not and hence, is not *Zuhanden*, "at hand," Heidegger emphasises the dynamic dimension, both verbal and adverbial, of the passing of the past. For the sake of clarity and conciseness, let us sum up the reasons why Ricoeur dwells on Heidegger's work in accordance with three points. First, the past is no longer conceivable as an entity that is immutable and can be tracked as such, in "its full availability." Second, the distinction between the past as that which is no longer and the existent state highlights the active relationship with the past, because individuals can delete or keep tracks of such past. Third, Ricoeur interprets Heidegger's analytic of existence as a theory of powers and non-powers, which is particularly fruitful for our analysis on legal memories and the right to be forgotten. In fact, the distinction between what is and what is not at hand (*Zuhanden*), has to do with that which remains available and thus can be erased, e.g. the deletion of memory traces. Therefore, from the stance of the remembering subject, the latter has to strike a balance between the poles of the full deletion and the total preservation of the memory traces. We return to this below.

In the opinion of Ricoeur, however, we should widen the perspective and insert the memories of the recalling subject "in the movement of exchange with the expectation of the future and the presence of the present": in other words, we should wonder about how memory and the past are connected with the present and the future (Ricoeur 1998). On the one hand, the present cannot be conceived as a mere result of the past, to which it would always be anchored. Rather, the present should be grasped in Kantian terms as the basis of the individual autonomy, namely as the ability to start something anew, to take the initiative to act on things in a non-predetermined way, that is, "to be free." From the legal point of view, this autonomy is what justifies the act of granting rights and duties to the individuals.

On the other hand, the role of memory involves the opening of the present to an unwritten future and therefore, according to Ricoeur, it also concerns the notions of fault and debt, much as the act of forgiveness, because "the orientation towards the future of the past is the counterpart of the opposite movement, for which the representation of the past affects the future." Think about the notion of fault, which is the burden that the past transmits to the future: whilst such burden weights on the future, the notion of forgiveness has to both clarify and release it. "If there is a duty to remember, it is because of guilt, which, by transferring the memory into the future, literally marks the future: you will remember! Do not forget!" (Ricoeur 1998). The representation of the past impacts on how we perceive the future and this is why Ricoeur suggests that a right memory is crucial, because the latter must account for the debt that the present has contracted with the past and, at times, dissolve the weight that nails the present down to a fixed and unchangeable past.

The twentieth century has imposed a duty to recall (Margalit 2002): more specifically, a duty to recall represents a form of justice, in order to not reiterate the crime by removing the memory of the victims. Still, today's advocates of the right to oblivion suggest an increasing need to be forgotten in this era of overwhelming memory, so as to be set free from the faults of the past and, so to speak, to be forgiven, for this would be the condition for a new beginning, a re-birth, and a new admission in the community. In light of Ricoeur's distinctions between "what is no longer" and "the existent state," between future, present, and the past, the connection between memory and oblivion that the idea of a right to be forgotten postulates should thus be deepened. Advocates and critics of the right to be forgotten usually discuss three points: (i) the construction of the personal identity; (ii) the relation between individual and collective memories; and, (iii) the different forms of oblivion vis-à-vis the idea of forgiveness. Our analysis proceeds with the first of such points, namely the structure of personal identities.

## 2.3 The Structure of Personal Identities

Advocates of the right to be forgotten often claim that a full control over data, which includes data removal (Mayer-Schönberger 2009), would guarantee the ability of the individuals to build their personal identity in a better way. After all, who could be in a better condition than the individual, so as to tell her own story, and give the latter a new beginning, a different articulation or, in more philosophical terms, a renewed understanding of the self?

Admittedly, this idea and its corollary seem legitimate and furthermore, we should admit that the information revolution makes the construction of personal identities through the control of data a critical issue today. Still, the structure of this argument has not to be taken for granted: what is at stake does not only concern a matter of personal identity, which depends on the link between past and future. Rather, we are dealing with issues of power and authority that have to do with the relationships that each of us has with other individuals. Therefore, the question correspondingly changes: how can the individual relate to her past and the others, without precluding the opening of the present to the future?

Again, Ricoeur's work and his interpretation of Heidegger's analytic of existence look particularly fruitful in order to tackle the issue. The indelible relation of the individuals with their past does not pre-determine or affect their lives but rather, represents the condition that allows the individuals to build an "inclusive" present and future. What constitutes both the essence and possibility of the future, according to Ricoeur, is our ability to understand and include the past in the dimension of the future and not the other way around, i.e., the dimension of the past that would always determine the future. The way in which we understand and include the past in light of the future ensures the maintenance of the self, namely of the individual that is incessantly transformed by the existence. This "inclusion" has been stressed by Heidegger in *Being and Time*: "it is truly be-coming only the Dasein that has authentically been" (Heidegger 2010). Although with hermetic language, the German philosopher means something simple: only that which comes from the past can aspire to have a future. Liberating the present from the tie that brings it back to the past may help people enjoying new beginnings, novel stories, or alternative lives, as advocates of the right to be forgotten often claim. Yet, the rupture with the past, e.g., the erasure of memory traces in the name of the right to oblivion, may deprive the individuals of their future in a subtle way. As Friedrich Nietzsche warns in his *Untimely Meditations*, namely in "On the Use and Abuse of History for Life" from 1874, it appears difficult to build our own future on a series of erasures, or removals. Indeed, the past irremediably affects that "maintenance of self" which allows the construction of every personal identity, notwithstanding the transformations of the subject. As seen in the previous section, such past should not be understood as something fixed and immutable, to be simply deleted or removed, but as something which is still incomplete and whose meaning is open to revisions or re-elaborations. This is an alternative way to build a story that frees the individual from the burden of the past: instead of erasing the memory traces, the aim should be to give them a new meaning.

However, the construction of such meaning, which is implicit in the reworking of the past, is not a simple individual or private activity. This construction needs several points of reference, e.g., the need that others have to access and know that past, so that every new meaning has to be shared with (and within) the community. This brings us back to the reasons why Ricoeur interprets Heidegger's analytic of existence as a theory of powers and non-powers: the power of reconstructing one's past deals with the power that others have to access and knowledge such past, which hinges as well on the sharing of knowledge that is essential for the formation and comprehension of a meaningful world. The power of the individual when relating to others must take into account these limitations that turn the previous condition into a non-power: our story—the story of ourselves on which we build our personal identity—never is a pure soliloquy but continuously draws on a common knowledge and a set of shared meanings. A personal story always depends on "the story of the others" (Ricoeur 1998): individual memories, as the attribution of meaning to the past, are not a private game but presuppose the social nature of language.

On this basis, the idea apparently suggested by supporters of the right to be forgotten, that memory is a purely private affair, based on the full availability of resources that feed the narrative construction of personal identities, is challenged. Whereas it is hard to sell the idea that a number of erasures, deletions, or disruptions of memory may set up the identity of the individuals, it also is difficult to conceive the reworking of the traces of the past and their meaning, on which the narrative of the self depends, as something that can be achieved regardless of the relationship with other individuals. As Ricoeur sums up, "the first fact, and the most important, is that you do not remember of your own, but with the help of other people's memories. In addition, our alleged memories often borrow stories heard from others. Finally, and this is the crucial point, our memories are framed by collective memories" (op. cit.).

The social nature of meaning makes problematic the widespread idea of the "ownership" of the past, according to which, once grasped the past as an entity, a trace, or a set of data, the past would fully be at the disposal of the individuals as the owners of their own memories. Since people's knowledge, memories, and meaning of the past are "irremediably social and public" (Ricoeur 1998), the right to be forgotten concerns how we should grasp the connection between individual and collective memories. This topic is examined separately in the next section.

## 2.4 Individual and Collective Memories

Advocates of the right to be forgotten often claim that the right to delete or remove the traces of the past, e.g., by subtracting them from the public dominion, allows the individuals to present themselves in a better light, consistent with the current and renewed image that such individuals aim to provide of themselves. Although this argument takes into account the connection between individual and collective memories, it may lead to a new paradox: by dissolving the tie which brings the present back to the past, the risk is to losing out on our own future. Whereas, in the previous section, attention was drawn to the meaning of the past, as such open to reinterpretations or revisions, here, following Ricoeur's reference to the work of Maurice Halbwachs (1992), focus is on how individual and collective memories are intertwined.

The individual memory cannot be understood as a purely private affair: such memory is not only supported by the memory of the subject but concerns social frameworks and external prostheses, such as speech, writing, signals, rituals, monuments, and shared organizations of space and time. Moreover, by describing the present from the selective reconstruction of the past, individuals define and negotiate their membership to the community, so as to make their present coherent with the group of which they are part. Contrary to the abstract representation of the individual as a self isolated from the rest of the community, every reformulation of the past has to do with a social task. The new meaning of the past shall be comprehensible in light of the conceptual framework with which every community builds its own collective memories. But, more concretely, how should we grasp the connection between individual and collective memories?

Ricoeur suggests paying attention to the work of Reinhart Koselleck (2004), and the distinction between the "space of experience" (*Erfahrungsraum*) and the "horizon of expectation" (*Erwartungshorizont*). Whilst the space of experience refers to the legacy of the past and its settled traces, the horizon of expectation is made up of all kinds of anticipations, such as desires, fears, or plans, that project us into the future. The idea is that of a "living present." Rather than the midpoint of a chronological chain, between a before and an after, the living present of the culture and the community of which we are part should be understood in light of this tension between the legacy of the past and the set of expectations that makes sense of "the dynamics of historical consciousness" (Ricoeur 1998). More particularly, in light of this tension, every reformulation of the past shows the power that the future exerts on the past. The individuals that intend to reinvent themselves through the deletion of the data ensured by the right to be forgotten, aim to make their story consistent with the set of expectations, desires, and fears, that form the living present of the community, or of the group to which they belong. The threat, often underestimated by the advocates of the right to oblivion, is that such alleged novelty, which the deletion of data and the rewriting of the past make possible, may paradoxically be flattened on the set of beliefs, desires, fears, and profiles that constitute the horizon of expectations in a given society.

On the one hand, by erasing the traces of the past, individuals may relocate themselves in the social fabric of the (new) group. Yet, on the other hand, there is the risk of maximum conformity to such group. The alleged autonomy of the individual, which the right to be forgotten intends to protect, can lead to the heteronomous clutch of the desire to conform to the horizon of expectation of a particular social group. Hence, the right to oblivion shows a hidden face, which is difficult to reconcile with the ideas of rupture, discontinuity, novelty, or rebirth, that the law should protect through the control over personal data. The reformulation of personal memories may go hand in hand with the consolidation of values, anticipations, and norms that ground the experience of which the horizon of expectation is made up. This consolidation is even more effective when presented under the banner of people's autonomy and the reaffirmation of the rights of individuals to freely reconstruct their own personality.

However, by insisting on how the articulation between individual and collective memories does not occur in a sort of social vacuum, it does not follow that a degree of oblivion is unnecessary for the elaboration of the past. Going back to Nietzsche's remarks in the *Untimely Meditations*, we may say that "it is possible to live almost without remembering, indeed, to live happily, as the beast demonstrates; however, it is generally completely impossible to live without forgetting." Yet, rather than a simple deletion of data, such oblivion should be grasped as a "plastic force … of growing in a different way out of oneself, of reshaping and incorporating the past and the foreign, of healing wounds, compensating for what has been lost, rebuilding shattered forms out of one's self" (Nietzsche 1874).

It is thus necessary to pay attention to the forms of oblivion and, hence, to a new argument that is commonly presented to support the right to be forgotten: the need to see the sins of the past forgiven, that is, to integrate the "wounds, lost parts, and shattered forms" into the fabric of the individual memory. Since forgiveness is the philosophical notion that refers to the human capacity to free the present from the weight of the past and open it to the future, next section deepens this figure, in order to determine its compatibility with the legal forms of oblivion and the right to be forgotten.

## 2.5 Political Oblivion and Forgiveness

According to Ricoeur, we should preliminarily distinguish between two forms of oblivion, namely between deep and manifest oblivion. The former concerns "memory as inscription, retention, preservation," the latter has to do with "memory as the function of recalling, of remembering" (Ricoeur 1998). Moreover, the notion of deep oblivion should be further distinguished between the forms of inexorable and immemorial oblivion. Inexorable oblivion "not only prevents the recall of memory … but works to erase the track of that which one has learned and experienced: it erodes the registration of memory as such" (op. cit.). The erasing of memory brought on by this form of forgetting deprives the holder of memories, much as any other person, of the possibility to access and know the past. This is the form of oblivion that conceives the past as *Vergangenheit*, namely that which is not, and that will not reoccur under any form in the present of memories.

Still, in the phrasing of Ricoeur, "there is the other pole of the deep oblivion, which would be better to define as the immemorial oblivion; it is the forgetting of the fundaments—of their original occurrence—that are not even 'events' of that which is possible to recall; we never really have learned them, but they make us what we are: the forces of life, the creative forces of history, 'origin', Ursprung" (op. cit.). The immemorial oblivion, in other words, is a condition of memory that deals with the past as *Gewesenheit*, rather than *Vergangenheit*, namely that which was and cannot be at hand any longer: the "being-state [that] makes oblivion the immemorial resource offered to the work of memory" (Ricoeur 1998). From this latter perspective, oblivion plays a positive role, because it makes manifest, rather than deep, oblivion possible: manifest oblivion does not concern the recognition, preservation, or deletion of the traces of the past, but the remembering through the evocation of memories. More particularly, Ricoeur distinguishes two different forms of manifest oblivion: passive and active oblivion. The former is the oblivion of removal or escape, through which individuals want to remove the painful memories of the past, or run away from them in advance. Passive oblivion, as a form of manifest oblivion, avoids inquiring and investigating, in order to break free from the weight of both individual and collective history, and its evil. As a sort of "I don't want to know," which does not entail a conscious activity but, more often, an ill-concealed form of negligence and omission, passive oblivion has characterized the collective memory of Europe in the past century (Darhendorf 1967; Friedlander 1992; etc.).

The other form of manifest oblivion, that is active oblivion, is selective and hinges on the strategic collection of memories. This latter choice is necessary, according to Ricoeur, at two different levels: the level of "life" and that of "narrative coherence." In the former case, it would be intolerable for any consciousness to recalling everything, or bearing the full weight of "the heavy load of the past." Likewise, in the case of narrative coherence, nobody can tell a story without omitting some events, episodes, or incidents, from the point of view of the plot chosen by the storyteller. From an evolutionary and pragmatic perspective, manifest oblivion, much as memory, have thus a crucial function, in that individuals and societies need to select what is deemed as significant, important, or useful, for the present. This evolutionary and pragmatic function shows why a politics of memory that reconciles remembrances and oblivion is necessary, since only their balance can have a "salutary effect" on human beings. As Nietzsche stresses in *On the Use and Abuse of History for Life*, "there is a line which divides the observable brightness from the unilluminated darkness, that we know how to forget at the right time just as well as we remember at the right time, that we feel with powerful instinct the time when we must perceive historically and when unhistorically. This is the specific principle which the reader is invited to consider: that for the health of a single individual, a people, and a culture the unhistorical and the historical are equally essential" (op. cit.).

In light of the fair balance that has to be struck between memory and oblivion, however, the focus of the analysis should be widened, so as to take into account forms of individual and institutional forgiveness, much as the ways in which the latter relates to the forms of manifest oblivion examined in this section. Whereas the notion of forgiveness can properly be conceived as opposed to any form of passive oblivion, forgiveness entails a sort of active forgetfulness, although, in the wording of Ricoeur, it "does not address the events in themselves, the trace of which must be carefully protected, but the guilt, whose weight paralyzes memory and, hence, the ability to creatively project ourselves into the future. The object of oblivion is not the past event, the criminal act, but its meaning and its place in the whole dialectic of historical consciousness" (Ricoeur 1998). As a result, forgiveness, as a form of active oblivion, does not intend to lose the tracks of the past but rather, aims to free the present from the burden that nails it down to a certain past, by changing the meaning of what happened.

Still, it belongs to the logic of forgiveness that is has to be asked for, so that individuals, who ask for forgiveness, are accordingly subject to the risk of rejection. This inter-subjective dimension of forgiveness, which represents the essential structure of the concept, brings us back to Heidegger's analytic of existence, interpreted by Ricoeur as a theory of powers and non-powers. Whilst the inter-subjective nature of the act of forgiving can help healing the wounds of memory, it also warns against "the ease of forgiveness. The claim that the practice of forgiving is a power, without passing the test of asking for mercy and, even worse, of rejecting forgiveness, triggers a number of traps" (Ricoeur 1998). In fact, as a sort of active oblivion, the will to forget one's fault, so as to free the present from the burden that nails it down to the past, i.e., the desire to be forgiven, cannot be conceived as a power but rather, as a request that must face the risk of rejection and, thus, as a non-power. Moreover, the inter-subjective nature of forgiveness cannot be equated with the relational structure of the law, e.g., the right to be forgotten, which can be represented as a subjective claim related to a position of duty. The mediation between the existential analytic of forgiveness as a non-power and current debate on the right to be forgotten as an individual power is given by the making of legal rules that find their legitimacy in a policy of "fair memory" and "difficult forgiveness" (Jankélévitch 2005). On one side, this memory policy has to be fair: as previously mentioned, forgiveness, as a form of active oblivion, aims to change the meaning of what happened, rather than erasing the traces of memory. This latter removal would simply deprive other individuals of the possibility to have access to the past, and may correspond to the mere desire to conform oneself to the "horizon of expectation" of the community. This aspiration, variously understood as a form of rebirth, of openness to the future, or novelty granted by forgiveness, is likely to flatten out on a tacit form of forgetting. The ambition of the autonomous subject to reinvent her own past often fits like hand to glove with the request to endorse the social norms and values of a given society.

On the other side, against the "ease of forgiveness," the memory policy should be difficult, both for philosophical and legal reasons. As to the former, in the phrasing of Ricoeur, "one should accept debts that are not paid, much as accepting either to be or remain an insolvent debtor, or that a loss persists. We should submit the guilty to the work of mourning, by admitting that the oblivion of escape and the never-ending persecution of debtors are the result of the same problem. One should draw a thin line between amnesia and the infinite debt" (Ricoeur 1998). Likewise, forgiveness should be difficult in the legal field, because the law has not only to transform an individual condition of non-power, related to the existential analytic of forgiveness, into the relational structure that connects a subjective right to a position of duty. Moreover, by protecting the informational identity of the individuals through the right to be forgotten, the latter has to be balanced against further rights, such as freedom of speech, of information, of press, etc. Recent case law mentioned in the introduction, e.g., the Court of Paris ordering some search engines to remove from their services all of the links that could trace an ex porno actress back to her previous life, the Italian Court of Cassation treating a newspaper as an online archive, down to the doubts of the Audiencia Nacional in Madrid, show the perverse effects that a misconceived right to be forgotten may have. The time is ripe for the conclusions of the paper.

## 2.6 Conclusions

In a world increasingly made up of data and information, where offline and online worlds converge, the need to protect people's personal data, the integrity of their informational identity and ultimately, the ability to tell their own story should not be underestimated (Floridi 2013). Legal instruments shall allow individuals to react and have a remedy for every violation of such rights. Yet, whether or not a right to be forgotten should be included among such rights is problematic: more specifically, today's debate revolves around the desirability of granting a right to oblivion, regardless of any violation of people's informational identity and the sphere of subjective claims, thereby conferring on individuals a general power to erase the traces of their past, so as to prevent that others may access and know it.

The paper has examined three arguments, supported by the advocates of the right to be forgotten, that *prima facie* appear intuitive, solid, and well founded. We summed them up according to the following points, namely: (i) the construction of personal identity via the revision of the past; (ii) the connection between individual and collective memories; and, (iii) the notion of forgiveness and both the need to forget the faults and to free the present from the burden that nails it down to the past. The common ground is represented by the individual aspiration to rebirth, to opening the present towards a different future, as a new beginning. By conferring on the individuals the ability to reinvent themselves through a new narrative, or a self-narration, the pact that binds the individuals to their community would thus be strengthened.

Still, the paper has insisted on the difficulties that affect this general framework. First, personal identities are not built from scratch but hinge on the dynamics of social relationships, through which common knowledge and shared meanings shape one's life experience. A re-birth, a new start, have to take into account this heritage of culture, language, experiences, etc. Second natures, as Nietzsche used to say, are weak constructions if they are simply built on rifts, breaks, and erasures. Second, the aim to reformulate old memories, so as to make them coherent with the horizon of the present, does not guarantee the autonomy of the individual self-narration. Rather, the desire for novelty may lead to the heteronomous grip of the social norms and values that define the community to which one belongs, in order to conform to the horizon of expectation that is part of the social space of the individual experience. Third, the desire to be forgiven and set the present free from the burden of the past, cannot be conceived as a power but rather, as a request that must face the risk of rejection. The novelty disclosed by the act of forgiving requires what Ricoeur calls an "additional working memory." Rather than removing the traces of the past, what is at stake here concerns an active form of oblivion, which is based on the policy of "just memory" and "difficult forgiveness." This is the twofold condition that justifies the subjective claim that connotes the right to be forgotten, related to the position of duty that others have in the legal system.

Finally, attention should be drawn to Ricoeur's remarks on the "status of documentary evidence." Time and again, this paper has stressed the inescapable, yet obvious, presence of other individuals as that which makes the right to oblivion problematic: memory is not a purely individual and private matter. The right to be forgotten cannot deprive others of their power to access (the knowledge of) the past, unless a specific violation of the individual informational identity is proven. Basic values and rights, such as freedom of information, depend on that access and knowledge. No society could survive without a public stored memory. In the phrasing of Ricoeur, "we can say that the memory is then stored: however, a stored memory ceases to be a memory in the proper sense of the term, in a relation of continuity and belonging to the present of consciousness, since it has access to the status of documentary evidence. Indeed, it is a trace to be followed and reconstructed by a historical consciousness but, first and foremost, it is a left trace, as with the passage of an animal. From this perspective, it already is a public entity. As an archived document, this additional status gives it a more institutional relevance, which corresponds to the qualified status of the profession of historians" (Ricoeur 1998).

The evocation of memories that move from a trace, rather than the present of consciousness, is thus linked to the events of a trace that was left. The latter, precisely because left, does not belong to someone any longer but becomes public and, hence, shareable. The erasure of the traces potentially limits such public space and the corresponding act of sharing, thereby curtailing such rights as the freedom of information and access, much as the process through which knowledge is formed, that is correlative to the institutional dimension of the trace. The protection of those freedoms and the formation of knowledge represent the condition for plural and even opposite narratives and, hence, a democratic society: "The comparison between conflicting forms of linking acts and events together can be supported by the firm pedagogical aim to learn how to tell our story from a point of view that is different from ours and that of our community. Telling otherwise, but also letting others to tell what we are" (Ricoeur 1998).

The institutional dimension of the trace and the idea of "telling otherwise" represent the other side of a coin in which forgiveness, as a form of active oblivion and non-power in Ricoeur's analytic of existence, brings us back to the problem of a "just memory." From a philosophical viewpoint, that coin should be weighed, so as to reflect on the dual status of the past and how individual and collective memories are intertwined. From a legal perspective, focus should be on how to strike the balance between the subjective claim to be forgotten and the protection of further rights such as freedom of speech, of press and, in general terms, the public access to knowledge and information. From a political stance, attention should be drawn to the mediation between the relational structure that characterizes the law in terms of claims and obligations, and the inter-subjective structure of oblivion that, following Nietzsche, could heal wounds, compensate for what has been lost, or rebuild shattered forms of the self. At the end of the day, today's debate has to match up with all these aspects of the right to be forgotten.

## References

Darhendorf, Ralf. 1967. *Society and democracy in Germany*. New York: Doubleday Garden City.
Durante, Massimo. 2011. Rethinking human identity in the age of autonomic computing: The philosophical idea of trace. In *Law, human agency and autonomic computing. The philosophy of law meets the philosophy of technology*, eds. M. Hildebrandt and A. Rouvroy, 85–103. London: Routledge.
Floridi, Luciano. 2013. *The ethics of information. Volume II of Principia Philosophiae Informationis*. Oxford: Oxford University Press.
Foster, Jonathan K. 2008. *Memory: a very short introduction*. Oxford: Oxford University Press.
Friedlander, Saul. 1992. *Probing the limits of representation*. Cambridge: Harvard University Press.
Halbwachs, Maurice. 1992. *On collective memory*. Chicago: The University of Chicago Press.
Heidegger, Martin. 2010. *Being and time*, (trans. by Joan Stambaugh, revised by Dennis J. Schmidt). Albany: State University of New York Press.
Jankélévitch, Vladimir. 2005. *Forgiveness*, (trans. by Andrew Kelly). Chicago: The University of Chicago Press.
Koselleck, Reinhardt. 2004. *Futures past: On the semantics of historical time*, (trans. by Keith Tribe). New York: Columbia University Press.
Margalit, Avishai. 2002. *The ethics of memory*. Cambridge: Harvard University Press.
Mayer-Schönberger, Viktor. 2009 *Delete: The virtue of forgetting in the digital age*. Princeton: Princeton University Press.
Nietzsche, Friedrich. 1874. On the use and abuse of history for life. In *Untimely meditations*, (trans. by Ian C. Johnston and partially amended by the Nietzsche Channel). https://webspace.utexas.edu/hcleaver/www/330T/350kPEENietzscheAbuseTableAll.pdf. Accessed 4 Nov 2013.
Pagallo, Ugo. 2011. ISPs & Rowdy web sites before the law: Should we change today's safe harbour clauses? *Philosophy and Technology* 24 (4):419–436.
Reed, Chris. 2012. *Making laws for cyberspace*. Oxford: Oxford University Press.
Ricoeur, Paul. 1998. La marque du passé. *Revue de Métaphysique et de Morale* 1:7–32.

# Chapter 3
# Location Data, Purpose Binding and Contextual Integrity: What's the Message?

**Mireille Hildebrandt**

**Abstract** This chapter investigates the issue of the proliferation of location data in the light of the ethical concept of contextual integrity and the legal concept of purpose binding. This involves an investigation of both concepts as side constraints on the free flow of information, entailing a balancing act between the civil liberties of individual citizens and the free flow of information. To tackle the issue the chapter starts from Floridi's proposition that 'communication means exchanging messages. So even the most elementary act of communication involves four elements: a sender, a receiver, a message, and a referent of the message' and his subsequent proposal that informational privacy can be described as 'the freedom from being the referent of a message'. After discussing the current environment of messaging in terms of Big Data Space and the Onl*ife* World, the chapter develops a more detailed definition for the right to informational *location* privacy. The road to this more detailed definition allows to highlight the balancing act inherent in both contextual integrity and purpose binding, and shows that the most salient challenge for such balancing acts is not—only—that Big Data Space and the Onl*ife* World turn contexts into moving targets. More importantly, the context of economic markets tends to colonize the framing of other contexts, thus also disrupting the

The research for this chapter was done in the context of the interdisciplinary research project on 'Contextual privacy and the proliferation of location data', funded by the Flemish Science Policy Agency (FWO), which entails a collaboration between computer engineers and lawyers from Vrije Universiteit Brussel and Katholieke Universiteit Leuven. I want to thank my co-researchers on the project for their many insights: Claudia Diaz, Laura Tielemans and Michael Herrmann. I also want to thank Helen Nissenbaum and Tal Zarsky for their comments on an earlier version of the paper and all participants to the 'Privacy Workshop: From Theory to Practice' at Haifa University in December 2013.

M. Hildebrandt (✉)
Institute of Computing and Information Sciences (iCIS), Radboud University, Nijmegen, The Netherlands
e-mail: hildebrandt@law.eur.nl

Jurisprudence, Erasmus School of Law (ELS), Rotterdam, The Netherlands

Law Science Technology & Society studies (LSTS), Vrije Universiteit, Brussels, The Netherlands

L. Floridi (ed.), *Protection of Information and the Right to Privacy – A New Equilibrium?*, Law, Governance and Technology Series 17, DOI 10.1007/978-3-319-05720-0_3,

protection offered by purpose binding. To safeguard informational privacy we need to engage in new types of boundary work between the contexts of e.g. health, politics, religion, work on the one hand, and the context of economic markets on the other. This ardent task should enable us to sustain legitimate expectations of what location messages are appropriate as well as lawful in a particular context.

## 3.1 Introduction: What's the Message?

Luciano Floridi has defined communication as 'exchanging messages'.[1] In this chapter I want to investigate whether this—cybernetic—starting point helps in understanding what EU legislation terms as 'data processing' and what Helen Nissenbaum has called 'flows of information'.[2] More specifically, I will investigate how it helps to understand the implications of the proliferation of location data for the right of informational privacy. Obviously, the sending and receiving of messages introduces actors that are not necessarily implied in the concept of data processing. Within EU jurisdiction data processing can refer to computational operations within computing systems, such as storing and retrieving of data or further manipulations such as data mining, that cannot be described as the sending of messages.[3] Even the collection of data is not necessarily a matter of senders and receivers, since the receiver of the data may collect it without any deliberate effort on the side of the data-holder. In fact, the data that was collected may have been 'manufactured' by the receiver, for instance in the case of clickstream behaviours or other types of machine-readable behavioural data. Thinking in terms of messaging clarifies this by highlighting that the data was not sent but taken, with our without consent. Phrasing the issue of informational privacy in terms messages also raises the question of what insights are gained (and lost) if we understand machine-to-machine communications as the exchange of messages instead of merely the exchange of data. The notion of a message seems to entail more than data, notably a direction and some form of—mindless or mindful—intent. This chapter aims to figure out how speaking of messages instead of data enhances or reduces our understanding of what is at stake with the proliferation of location data.

As Floridi notes, 'even the most elementary act of communication involves four elements: a sender, a receiver, a message, and a referent of the message'.[4] This highlights the flow of messages between senders and receivers, thus qualifying the notion of information flows in terms of points of departure and arrival and specifying the 'aboutness' of the information in terms of a referent (rather then, for instance,

[1] Cf. the introductory text for the Workshop 'Protection of Information and the Right to Privacy: A New Equilibrium?', held on 21 June 2013 at the European University Institute, Fiesole, Italy.

[2] On cybernetics Wiener (1948). On EU data protection De Hert and Gutwirth (2006). On contextual integrity Nissenbaum (2010).

[3] In intra-machine data processing the actors could be those ordering or operating the processing; in the EU legal framework these actors are defined as data controller and data processor. From the perspective of cybernetics the actors would be the machines (software and/or device) that send the message, referring to a mindless form of agency.

[4] See n 1.

an owner of the data). In respect of the proliferation of location data, building on Floridi,[5] we can now formulate four fundamental rights and freedoms that concern location data in terms of the exchange of messages: (1) freedom of speech concerns the right to send messages from whatever location to whatever location, including the right not to be located when exercising freedom of speech, (2) freedom of information concerns the right to receive messages from whatever location to whatever location, including the right not to be located when exercising freedom of information, (3) communication security concerns the right to protection from unwanted access to one's location data, manipulation of one's location data or destruction of one's location data (confidentiality, integrity and availability CIA) and (4) the right to informational location privacy concerns the freedom from the referent's location data being shared without consent or necessity.[6] Though all these rights and freedoms can thus be translated into location-data-relevant formulations, this chapter will limit itself to informational privacy in the broad sense of what the OECD has called the 'fair information principles (or policies)' and what is defined as 'data protection' within the European Union (EU). In saying that informational privacy refers to the requirement that information is shared on the basis of either consent or necessity I hope to catch both purpose binding and contextual integrity as normative frameworks that delimit (1) data processing, (2) flows of information, (3) the sending of messages.

In the following sections I will first discuss informational location privacy in the context of Big Data Space, followed by the introduction of three types of data with a large impact on autonomy, identity and privacy: volunteered, observed and inferred data. This results in articulating informational location privacy as relating to 'raw', networked and 'processed' data. The impact of this data will be further explained and developed by investigating the consequences of various types of location messages in the context of the so-called Onli*fe* World, challenging traditional (modern) notions of autonomy,[7] identity and privacy. All this should create the middle ground for the sections on the ethical concept of contextual integrity, notably the contextual privacy decision heuristic, and the legal obligation of purpose binding as exemplified in the EU framework of data protection. Finally then, I will evaluate how framing informational location privacy in terms of messaging helps to understand to what extent contextual integrity and purpose binding are side constraints or require a balancing act.

[5] See n 1.

[6] Communication security is the odd one out, since it is not a fundamental right. One can, however, easily relate it to the foundational tasks of the state in securing critical infrastructure, and safety and/or relate it to the confidentiality of communication that is at stake in the right to informational privacy.

[7] Obviously modernity constitutes a tradition, despite the fact that it is often framed as liberating itself from any type of tradition.

## 3.2 A Cybernetic Starting Point: Location Data in *Big Data Space*

The idea that an act of communication can be defined as the exchange of messages takes its clue from Wiener's theory of cybernetics. Wiener connected the notion of communication with that of control, claiming that the exchange of messages is meant to give agents a certain measure of control over their environment. He formulated his theory to explain communication between machines, explicitly defining human persons as machines. In doing so, he hoped to enhance scientific understanding of human-to-human, human-to-machine and machine-to-machine communication. Though we need not agree that human persons are machines, it makes sense to follow up on Wiener's semantics for the simple reason that our online and offline environments are increasingly constructed and 'animated' by interactive computing networks, built on the semantic assumptions of cybernetics.[8] By adopting the idea that communication is a matter of messages sent and received, with a content that refers to something outside of the message, we can for instance flesh out to what extent human persons are indeed messaging machines and, if so, to what extent their messages differ from those of other messaging machines (plants, animals, robots, artificial agents).

Beresford and Stafano have defined location privacy as 'the ability to prevent other parties from learning one's current or past location'.[9] Based on Floridi, we can translate this as 'the freedom from the location of the referent, the sender or the receiver of a message being shared'. This highlights the idea that privacy is a liberty (*freedom from sharing*) rather than an issue of control (*the freedom not to share*). Obviously, Beresford and Stafano emphasize the aspect of self-determination in the narrow sense of control (the freedom to share or not to share). However, defining informational location privacy in this way is too absolute. We need to take into account that informational privacy does not equate with hiding per se, but with the *capability* to hide or remain hidden *if there is no necessity or consent for sharing information.* Note that the EU legal framework is focused on discrete *personal data*, whereas my definition is focused on a *data flow*.[10] Similarly, as noted above, the EU legal framework is focused on the *processing* of personal data, which includes all kinds of operations such as the recording, storing, retrieving, computing, deleting, pseudonomysing or anonymising of personal data, whereas my definition is focused on the *sending of messages containing* personal data. The advantage of thinking in terms of data flows and messages could be that it gives prominence to the dynamic and interactive character of exchanges of location data. This does not

[8] With a semantic assumption I mean an implicit understanding of the meaning of the foundational concepts of cybernetics. The fact that cybernetics is more interested in syntaxis than in semantics obviously does not entail that its own vocabulary is devoid of meaning. On the history of cybernetics Hayles (1999).

[9] Beresford and Stajano 2003.

[10] This is also one of the important advantages of Nissenbaum's understanding of contextual integrity in terms of information flows, see Sect. 5 below.

imply that the processing of location data that is performed by computing systems is of no relevance, but it allows to discriminate between intra-machine processing on the one hand and the exchanges of 'processed' location data between machines and humans on the other. Especially when specific decisions are taken on the basis of 'processed' location data, it is important to distinguish the processing from the exchange, and both from the decisions they nourish. In this chapter I will, therefor, use the term 'processed' location data as referring to location data that have been 'refined' by computing systems that use 'raw' location data as a resource for what some have called 'data derivatives'.[11] Because it is possible to infer location data from other data (e.g. from mobility patterns or energy usage behaviours), I will also use the term 'processed' location data for inferred location data. Though EU data protection law uses a broader definition of data processing, I want to discriminate between the first 'making' of the data and the various products 'made' by further processing of the initial data. This highlights the difference between 'raw' and 'inferred' location data on the one hand, and between 'raw' location data and inferences drawn from such location data about other aspects of an individual person on the other hand.

Before further exploring the particulars of location data in terms of a message, we need to discuss the environment in which all this communication takes place. To do this I will introduce two new terms: the first being *Big Data Space*, the second the *Onlife World*. Both terms highlight, first, the computational layers which constitute large parts of our environments, and, second, the hyperconnectivity of the emerging life world. Thirdly, they foreground the increasing entanglement of online and offline environments. In this section I will focus on the notion of Big Data Space, leaving the discussion of the *Onlife World* for Sect. 4.

*Big Data Space* refers to the fact that the amount of available data enables new types of artificial intelligence, notably knowledge discovery in databases (KDD) and machine learning (ML). Paraphrasing Mayer-Schonberger & Cukier we could say that Big Data enables to do things that are not possible with 'small data'; Big Data introduces differences that make a difference,[12] though we may not yet be in the clear on what difference is crucial. *Big Data Space* also refers to the fact that databases are fused or matched, while the knowledge that is inferred can be stored, sold and re-used in other databases, thus generating a network of interconnected data servers, inference machines and virtual machines that constitute a complex, textured space with distributed access points.[13] To the extent that this space is connected with the Internet we can call it cyberspace, but since many interconnected computing systems are not connected with the Internet (various types of 'walled

[11] The term 'data derivatives' was coined by Amoore (2011). With raw data I do not mean to suggest that 'data' is somehow 'out there', merely to be picked up. Digital data is always a translation from the flux of life and already incorporate specific assumptions about what experiences or observations qualify as what type of data. See Gitelman (2013).

[12] On these differences see Mayer-Schönberger and Cukier (2013), boyd and Crawford (2011), Hildebrandt (2013a).

[13] This also refers to cloud computing, which changes the scope, the security, the availability, accessibility, the distribution and the virtuality of the space of and for Big Data.

gardens' like the NSA and data brokers like Axciom or Experion) I will speak of *Big Data Space*, taking note of the fact that this is neither a homogeneous space nor a space that can be defined in purely spatial metaphors. *Big Data Space* is a timespace that synchronizes data exchanges, involves massive parallel processing, and challenges traditional notions of past and future. It combines an external memory for text, images, computing programs and real-time pattern recognition with a plethora of techniques for predictive analytics and feedback mechanisms. Other than the external memory constituted by written and printed text *Big Data Space* is radically dynamic and polymorphous, while its operations are informed by complexity, because they are, to some extent, recursive—due to the use of ML techniques that persistently *nourish on* and *reconfigure* the timespace of Big Data.

Taking into account that location data will often be situated in *Big Data Space* we must acknowledge that location data will often be 'processed' data and/or networked data. The latter is data that is or can easily be linked with other location data of the same person, of other persons, or with other types of data (e.g. purchasing data, energy usage data, video consumption data, education data, employment data). Informational location privacy should therefor include 'the freedom from networked and/or 'processed' location data being shared with others without consent or necessity'. Especially when decisions are taken about the referent, sender or receiver that are based on networked and/or 'processed' location data, achieving location privacy would imply that such data have been shared with informed consent or based on the necessity required by, for instance, EU data protection legislation. Note that location privacy is not about hiding or controlling one's location data, but about the conditions that must be met when location data and its derivatives is being shared. Note, also, that these conditions are not formulated as balancing acts but as side constraints; if they are not met the sharing is unlawful. That is why informational location privacy is a freedom: the freedom from unlawful sharing of 'processed' or networked location data. Such a formulation does not preclude that a justifiable interpretation of the side constraints may involve a balancing act, notably the proportionality test that is inherent in the condition of necessity.[14]

Lawyers will focus on location data that relate to an identified or identifiable natural person, because this constitutes 'personal data' (in the EU) or 'personally identifiable information' (PII, in the US).[15] However, 'processed' location data may consist of patterns or inferences that do not qualify as personal data, though they affect a person whose location data matches such 'processed' location data. This raises the issue of whether the concept of personal data or PII is salient or even adequate in addressing the notion of informational location privacy. In the next section we will follow the trail of the World Economic Forum (WEF) that has made the attempt of rethinking personal data in the era of Big Data Space, taking into account that the monetization of personal data is the driver for a number of business models. In distinguishing volunteered, observed and inferred data they propose to develop a more refined understanding of what is at stake in the era of Big Data, KDD and ML.

[14] See the last section of this chapter.

[15] PII and personal data are defined slightly differently and the legal effect of a data being qualified as PII in the US or as personal data in the EU differs.

## 3.3 Three Types of Location Data

The common sense on informational privacy seems strongly attached to the idea that personal data are owned by the person to whom they refer, leaving it up to her to decide to either share or hide them. The notion of ownership is confusing here, because it implies that we are talking about an exclusive right to a rivalrous good. A rivalrous good cannot be possessed by more than one person: once I take it from you, you don't have it anymore.[16] Personal data does not fit that category; it can easily be shared with a number of people without taking it away from whoever it refers to. In fact, many 'processed' data were never in the possession or even awareness of their referent. Indeed the whole idea of personal data is to share information about oneself, to allow others to identity and address one. If you get to know my name you may have little use for it if everyone else—including me—is forced to forget it (this would be the case if it were a rivalrous good). The fact that personal data is often discussed in terms of ownership is of course related to the fact that people feel strongly about the knowledge and information that concerns them, which they believe to *belong to* them. This leads people to claim that they are somehow *entitled* to it. Such entitlement, however, does not imply exclusiveness. Different legal subjects can have different types of entitlements to the same data. Take, for example, energy usage data. If it refers to a particular household with identifiable users, energy usage data is personal data for those that are capable of linking the data to the identity of the user. The energy supplier that has a contract with the user will need subscription data to prepare the bill and location data to supply the energy. This means that the user has data protection rights towards the supplier, while the supplier has the right to require, store and retrieve the data for the purposes of energy supply and billing. Obviously, based on energy usage patterns, the supplier could *infer location data* for the members of the household based on their use of electricity or gas. To the extent that the supplier has no need for such data, a supplier that is active within the EU jurisdiction is not allowed to process them. The purpose binding principle stipulates that personal data may only be processed for an explicit, specific and legitimate purpose and may not be reused for an incompatible purpose.[17] So, informational location privacy here means that energy suppliers should refrain from sharing 'processed' location data of their subscribers.

Recently, the WEF has been discussing the tensions between the interests of senders, referents, recipients, users and processors of data in terms of volunteered, observed and inferred data.[18] This may enable a more precise understanding of what is at stake with the proliferation of location data. *Volunteered data* are the data

[16] Joint possession of the same good is possible, but that is not the point here. On property of personal data see e.g. Prins (2006) and Purtova (2012).

[17] See art. 6 of the Data Protection Directive (DPD): D 95/46/EC and art. 7 of the draft General Data Protection Regulation (dGDPR). To supply energy and to address the bill, the supplier must have the location of the household; it is, however, not allowed to infer and use the location of individual persons within the household for other purposes than energy supply and billing.

[18] World Economc Forum (2011, 2012).

that people deliberately provide, and often also 'make':[19] pictures or text posted on Facebook, emails sent to friends or colleagues, credit card details or an address for the delivery of a book. Volunteered data are part of a message, sent by the referent of the message to a particular or even to an unlimited audience (e.g. in the case of a publicly accessible blog). Social Networking Sites (SNSs) such as Foursquare enable people to share their location data with friends, to make themselves visible and reachable in a certain location. Users of Foursquare basically *have* messages *sent* on their whereabouts. However, Foursquare may decide to retain the content of the message and its metadata in order to sell such data to providers of location-based services or personalized advertising. This is not because the user had sent a message to Foursquare, but because Foursquare *observed* the location and 'datafied' it to enhance its business model.[20] Datafication refers to the process of translating the flux of life into discrete, machine-readable data points. The message sent by the user was intended for her friends; the SNS was merely the enabler. However, as we all know, the enabler makes its money by using the behavioural data (of which location is but one) to pay for its operations and to make a profit. So, the location data is both volunteered (in regard the friends) and observed (by the SNS). It is important to acknowledge that whether a data is volunteered or observed depends on the relationship between sender and receiver and not on the data itself; this implies that the same data may be volunteered within the relationship between the user of an SNS and her friends, but observed within the relationship between the user and the SNS provider. On many occasions the entities that collect behavioural data do not even have a relationship with the person whose behaviours are datafied. Advertising networks such as Double Click (now Google) and services such as Google Analytics (guess what, also Google) are employed by web portals, web shops, and a host of online service providers (whether public or private), who observe and 'process' online behavioural data on behalf of whoever wants to 'improve the user experience' or their own profit (which is assumed by some to coincide). For instance, websites often employ so-called A/B research design to personalize their interface to the observed or inferred location of the visitor, e.g. by adapting the language, the currency and—of course—the price of the services that are offered. The relationship between the user of the SNS and those third parties can be qualified as eavesdropping, if we think in terms of messaging.

Observed data are the measurable behaviours of 'onliners' and 'offliners' that can somehow be datified: click stream behaviours online, transaction behaviours that involve loyalty cards, public transport behaviours read from the public transport smartcards, health related behaviours that feed into remote healthcare systems,

[19] To the extent that such data form a 'work' by an 'author' they generate copyright.

[20] Datafication will also generate copyright or other intellectual rights, but now on the side of the service provider (the observer) of the data. Whether this is the case depends on the jurisdiction and the nature of the process of dataficiation. For instance, a patent on the software that creates the data may be copyrighted or patented, the database that is used to store the data may entail a sui generis IP right or a copyright, the data mining software may be patented or subject to copyright. An interesting question is whether the data itself is the object of an IP right on the side of the 'data-fabricator' or whether it can claim be subject to protection as a trade secret.

traffic data of telecom end-users and the more. These data are not necessarily volunteered: they need not be deliberately provided or fabricated by the person whose behaviours they refer to. They are 'made', 'constructed', 'read', 'measured' by a plethora of computational machines that are increasingly adapting online and offline (our Onli*fe*) environments to suit inferred preferences of the user (or of whoever pays for them). *Big Data Space* is stuffed with observed data; i.e. with datafied behaviours of individuals, crowds, eye-movements, weather conditions, products (life cycle management), skin conditions, eye-movements, gait, financial transactions, security vulnerabilities, blood composition, whatever. Critical infrastructure is increasingly dependent on such observed data (cf. the smart grid) and most business models cannot gain competitive advantage without them. Even our governments display a firm belief in the added value of massive datafication (think NSA, but also China or Europe—each in its own way, with its own justifications). Location is an easy target for datafication in the era of smartphones and other mobile devices, products enhanced with radio frequency identification (RFID) tags, CCTV camera's and other gear that enables to locate an individual person in timespace. Apart from location based services (LBSs) most of the datafication will concern observed location data.

The added value of volunteered or observed location data is not so much in the growing aggregation of discrete data points, even if these are traded and monetised in the high frequency markets of advertising space or stored for as yet unforeseen future re-use. The added value is in the inferences. Here we encounter the most interesting privacy paradox. Volunteered and observed data will often be personal data (insofar as they relate to an identifiable person), whereas inferred data concerns patterns and correlations at a higher level of (statistical) abstraction that cannot be qualified as personal data. However, it is precisely these patterns that form the trove against which our data points are matched and correlated. The inferred data are the gold that is mined from the 'raw' (volunteered or observed) and the 'processed' or networked location data. Not only do these inferred data have a more permanent and transformative impact in the Onli*fe* World, they lack the protection available for 'unprocessed' data while they will often enjoy protection as part of the trade secret or intellectual property rights of those who invested in producing them.[21]

Volunteered data clearly constitutes messages, intended for one or more specific addressees. It may, however, be received by other parties that observe such data to enhance their business case. Though observed data may also be defined as constituting a message, it is not entirely clear what is the meaning of a 'sender' in that case. On top of that we need an extra term to distinguish the addressee of the message from the receiver (though they may coincide). One way of analysing observed data as a message is to qualify the machine, the software and/or the hardware) that enables observation, as the sender. Another way would be to qualify the receiver as the sender to the extent that the receiver has initiated the process of having the data sent to its own processing engines (e.g. by means of cookies, or browser fingerprinting). Finally, one could simply say that the data is taken instead of being

[21] See e.g. recital 42 of the current Data Protection Directive D 95/46/EC.

sent; by highlighting that no message was sent while data was still captured the difference with volunteered data stands out. To describe observed data in terms of messaging we seem to require the concept of intent, raising two further issues: first, does sending imply intent?, and, second, should we accept the notion of 'mindless' intent to refer to machine-to-machine exchanges of data? I will leave this in the middle for now, and conclude that whereas volunteered and observed data can both be understood as messages, inferred data is another matter. Data derivatives may form the content of a message, but—like other types of 'processed' and networked data—they do not necessarily involve a data exchange. In the next section I will discuss volunteered, observed and inferred location data in the context of the emerging Onl*ife* World, hoping to flesh out how the messaging of 'raw', networked and 'processed' data is impacting everyday life in this new 'Onl*ife* World'. This should create a middle ground to discuss informational location privacy in terms of contextual integrity and in terms of purpose binding.

## 3.4 Beyond Cybernetics: Location Data in the *Onlife* World

The 'Onl*ife* World' is a concept developed by the Onl*ife* Initiative, a group of philosophers, social scientists and researchers of artificial intelligence, brought together by Nicole Dewandre and Luciano Floridi.[22] The aim has been to contribute to the reengineering of current conceptual frameworks. Though concepts cannot be 'fixed' in a mechanical way, they can be in need of mending or even reinvention. It should be clear that traditional (i.e. modern) conceptions of self, mind and society have been disrupted by the rapid transformations brought about by game changers such as the mobile smartphone, algorithmic search engines and online social networking sites. In speaking of conceptual reengineering we refer, for instance, to the notion of philosophical engineering as used by one of the founding fathers of the world wide web, Tim Berners-Lee, who exclaimed in an email exchange: '(…) we are not analyzing a world, we are building it. We are not experimental philosophers, we are philosophical engineers.'[23] I read this as a call for awareness, addressing those who engineer the information and communication infrastructures

[22] See < https://ec.europa.eu/digital-agenda/en/onlife-initiative >. I am one of those 'gathered' by the initiators and the many in-depth discussions have inspired my own thinking, especially complementing my research into the computational turn with more focused attention to the hyperconnectivity of the emerging lifeworld.

[23] See http://lists.w3.org/Archives/Public/www-tag/2003Jul/0158.html; Hildebrandt (2013b), p. 235. Conceptual engineering can also be understood as derived from Carnap's logical positivism, which aimed for an '*unphilosophical philosophy*, (…) building up from clear, technical, first principles. (…) striving for 'a "modern" way of life, (…) grounded on a vision of the machine age' Galison (1990), p. 750. My own link with philosophical engineering hooks up with Tim Berners-Lee's exclamation that engineers are constructing and shaping our lifeworld. I take a pragmatic and phenomenological perspective, cf. Ihde (2008).

of our current era, reminding them of the constitutive impact of their building, crafting and tinkering on what can make or break us as individuals, as societies, and as increasingly onl*ife* hybrids. For me, the concept of an Onl*ife* World tweaks the increasingly inadequate notions of online and offline, while focusing on what this means for our 'lifeworld', in both the everyday and the phenomenological sense of the term.[24] The Onl*ife* Initiative thus admits that some of the foundational concepts of modernity are inadequate, insofar as they are incapable of coping with the relational nature of the self and the increasing heteronomy of human-machine relationships. The intuition that triggers the Initiative is that both hyperconnectivity and invisible computational decision systems challenge vested notions of, first, human autonomy; second, Westphalian sovereignty and; third, the common sense difference between mind and matter. *Big Data Space* enables pattern-recognition that allows for subliminal manipulations of consumer preferences that correlate with 'raw', networked and 'processed' location data, thus challenging the assumption of human autonomy; it sparkles cross-border access to 'raw' and 'processed' location data by law enforcement and foreign intelligence services, thus challenging the assumptions of internal and external sovereignty; and, finally, *Big Data Space* enables computing systems to develop of a mind of their own—acting on the feedback they infer from their environments, thus challenging the experiential duality of passive matter versus active mind. The latter is especially relevant with regard to location data, since smart environments may confront individual persons with anticipations of their 'when whereabouts'.

At the same time, we are confronted with the experience of hyperconnectivity—across the extended timespace of messaging services such as e.g. skype, sms, WhatsApp, email, and across the hyperlinked virtual space of the World Wide Web, the page rank algorithms of search engines and the scaling of interrelationships in the realms of social networking sites. This entangles us with the network effect of complex non-linear relationships of cause and effect. A fundamental unpredictability has surfaced, leaving us with a sense of uncertainty and liquidity; presenting a trove of surprising opportunities (novel business models, scientific discovery, risk management) and devastating misfortunes (e.g. the financial crisis). Such unpredictability changes the meaning of meaning, disrupting the foreseeability of the consequences of our actions, thus reducing or even transforming our understanding of human autonomy and undoing the assumptions of national and international jurisdiction, while making us dependent on the technological infrastructures that mediate and constitute our environment.

The philosophical concept of the lifeworld, coined by Husserl and further developed by phenomenologists such as e.g. Merleau-Ponty, Ricoeur, Varela and Ihde,[25] refers to the way we perceive, cognize and co-constitute our environment, while at the same time configuring our sense of self and society. It regards the way we are 'at home' in the world, navigating familiar surroundings, anticipating the habits and habitations of our fellows and of the institutions or social structures that

[24] Husserl (1970); Ihde (1990).

[25] Merleau-Ponty (1945); Ricoeur (1976); Varela et al. (1991); Ihde (1990).

co-determine our consolidated expectations. Philosophers of technology, such as Ihde and, for instance, Verbeek,[26] have highlighted the enabling as well as constraining role of technologies and technological infrastructures (the script, the printing press, mass media, hyperlinked connectivity and computational in-betweens) in the co-constitution of self and lifeworld. Ihde and Verbeek speak of technology in terms of mediation, emphasizing that such meditation entails different types of impacts on how self, mind and society are shaped. The introduction of the handwritten manuscript reconfigured our relationship to time and space; it enabled a distantiation between author and reader across geographically distant lands and between temporally distant eras. In a way, it liberated human beings from the tyranny of the here and now that prevails in face-to-face relations. Location was multiplied by imagined and remembered locations beyond the memory and forecasts of individual human minds. The externalization of memory has created both history and—paradoxically—a plethora of present futures that co-constitute the future present.[27] Computational mediations by what Greenfield has called the 'everyware',[28] reshuffle our connections to the locations we inhabit and those we visit, either 'in the flesh', electronically or virtually. As Julie Cohen has explained,[29] our sense of location multiplies: our embodied self sits behind a screen, while communicating via email, posting messages on SNSs, or while engaging in real time interactions in online gaming, video conferences and the more. Note that, currently, we have not the faintest idea of where the physical servers are *located* that allow us to send and receive messages, though we can no longer assume that they remain within the confines of a jurisdiction we know well enough to trust. Location matters, but its datafication uproots traditional properties of 'place' as a coordinate that is independent from 'time'.

How does our cybernetic point of departure relate to the Onl*ife* World? Thinking in terms of messages has the advantage of paying attention to the flow of information, while also taking into account that data can only mean something to *somebody*—data in itself is not just mindless but also meaningless. Viewing data as moving in a specific direction, from a sender to a receiver, enables to see data as content in the context of a specific relationship. Moving beyond the cybernetic focus on the *integrity* of the data that is 'transported' from one machine to another,[30] we can instead ask the question whether the same data means different things to the sender and to the receiver, and, if so, on what this depends. Is meaning agent dependent? If so, how can the agent-sender foresee how her message is understood by the agent-receiver? Can she tune her message in a way that increases the likelihood that the addressee gets the message that she is trying to convey? Might this depend on the

[26] Verbeek (2006).

[27] Cf. Esposito (2011).

[28] Greenfield (2006).

[29] Cohen (2007).

[30] The integrity refers to the fact that the content of the message remains the same during the exchange. A similar focus is present in digital security: next to confidentiality and availability of data and systems, digital security is focused on making sure that the data sent is identical with the data received.

role of the agent-addressee, and thereby on the context within which the message will be received? This connects with what was briefly discussed above, namely that a sender may intend to send a message to a specific addressee, whereas the message is (also) received by one or more others.[31] As indicated, this introduces the notion of intent and raises the question of whether speaking of messages implies agency and what this means in the context of machine to machine messaging. These questions gain traction in an Onl*ife* World that is defined by the hidden complexity of vast layers of computational in-betweens and by the network effects of hyperconnectivity. How do agency, intent and the difference between addressee and receiver relate to informational location privacy in the Onl*ife* World? Does the emergence of an Onl*ife* environment afford something like 'the freedom from 'raw', networked and/or 'processed' location data being shared with others without consent or necessity'? Or should we acknowledge that the mindless agency of machine to machine communication renders both consent and necessity meaningless as effective constraints on the sharing of information? In the following sections I will investigate how informational location privacy defined in terms of the sending of messages, relates to the ethical concept of contextual integrity and the legal concept of purpose binding.

## 3.5 The Ethical Concept of Contextual Integrity

Informational location privacy implies that location matters to individual persons and relates to a sphere that requires boundary work.[32] The right to privacy is often defined in relation to the sanctity of the home as a physical location that shields the person from outside interference. To put it bluntly, this is the sphere where one can burb and scratch, get up late or sit through the night, eat, dance, read, drink and watch television without being supervised. Whereas we may wish to portray a certain image of ourselves when going off to work, visiting one's parents-in-law or when we walk the streets of an unknown city, the home provides for a space of retreat, of freedom from external constraints, from the gaze of the other and from the investigative powers of both one's neighbours, the family and the state. I hope that the reader will detect a certain irony here, since the state has found its way into our homes via e.g. the interception of telecommunication; family is often—a potentially oppressing—part of the home environment; and neighbours can violate our sense of privacy by means of e.g. loud music or gossip. Nevertheless, the matter of walls, doors and windows indicates a solid, visible and durable kind of boundary work that differs from much of the boundary work required in an Onl*ife* World, where the borders between work, home and leisure have to be built into email traffic, facebook friending strategies and online websurf and purchasing behaviour

[31] This is core to digital security: it relates to the confidentiality and is usually discussed in reference to Alice sending a message to Bob, while Eve is evesdropping on them to overhear confidential information. See Leeuw and Bergstra (2007), and—just for fun: Gordon (1984).

[32] On privacy as boundary work rather than control Altman (1975).

patterns.[33] Simple oppositions such as private and public seem to lose their meaning in a world that sets the defaults for seamless bordercrossing between a host of different social spheres, allowing but also forcing onli*f*ers to continuously navigate the furiously overlapping contexts of e.g. employment, business, consumption, religion, health, family, politics and education. Navigating these 'furiously overlapping contexts' must take into account that messages sent within one context will often—though unintended—arrive in another context, notably due to the fact that most of these messages concern observed data instead of—or next to—volunteered data. Framed in another way, much observed data is 'gleaned', even though no message was sent.

This raises the issue of context. In her ground-breaking work on the ethics of data sharing, Helen Nissenbaum has called for a more nuanced, more thoughtfull but also more practical understanding of what is at stake with informational privacy. After publishing pivotal work on 'privacy in public' and 'contextual integrity'—besides numerous other work e.g. on trust and security, Nissenbaum has expounded on the idea of privacy in context, explaining how we might rethink the integrity of social life.[34] In this section I want to explore how a cybernetic understanding of the right to privacy can be transformed by the broader scope and enhanced by the more precise articulation made possible by the introduction of the concept of contextual integrity. This, however, does not mean that a cybernetic understanding of privacy in itself brings no added value or can be discarded as merely reductive. As mentioned above, I believe that it is crucial to develop and operationalize conceptions of informational privacy that are interoperable with their cybernetic articulation, precisely because our Onli*f*e World is saturated with computational systems built on cybernetic assumptions.

Nissenbaum defines contexts as structured social settings, with characteristics that have evolved over time.[35] They are subject to a host of causes and contingencies of purpose, place, culture, historical accident, and the more. In traditional sociological and philosophical terms one could say that a context is an institution, a social sphere, a practice, entailing roles and patterns of interaction. Some of the examples she gives are health care, employment, education, religion, family and the commercial market place. Different contexts may overlap or conflict, one context may 'nest' in another. In fact, I will argue that one of the most forceful challenges for contextual integrity occurs when one context monopolizes a specific domain or even an entire society (e.g. the context of religion may dominate the private sphere or even the political sphere as in a theocracy), or that one context colonizes another (e.g. the commercial market place may colonize higher education). Importantly,

---

[33] The introduction of personal computing and smart phones has blurred the borders between home, work and leisure, while it has enabled detailed monitoring of web surf behaviours that renders transparent one's personal preferences. On top of that, smart energy metering systems allow to detect unexpectedly granular lifestyle patterns, potentially providing an x-ray of what goes on within the home.

[34] Nissenbaum (2010).

[35] Nissenbaum (2010), p. 130 ff.

Nissenbaum suggests that context is not a formally defined construct, it cannot be represented in a final, definitive way. This does not mean that nothing can be said about what counts as a particular context, but one should always take into account that contexts are constituted by people and norms that are co-constituted by the contexts they navigate. Context is—I would suggest—firmly grounded in the thin air of our double contingency;[36] contexts *make us up* while we *make them up*. That being said, for individual persons the norms that constitute and regulate particular contexts are mostly given, even if they may find ways to challenge, test or transform them.[37] It may be, however, that this experiential fact—that we are somehow thrown into an already existing socially structured world—is less obvious than before. It seems that, first, the blurring of borders between different contexts and, second, the fact that a person can easily navigate different contexts from one location, has a lasting effect on the stability of contexts. The point is that contexts have to tune their song to the constant interference of competitive contexts that impose themselves and vie for our attention. When arguing for contextual integrity we should therefor acknowledge, first, that context is becoming a moving target and, second, that we are confronting a power play between the contexts of—notably—the political and the economic spheres on the one hand and the spheres of healthcare, education, employment and religion on the other hand. Populism and market fundamentalism may overrule common sense understandings of what matters in a healthcare or employment context and this raises the question of what contextual integrity means in terms of data flows.

Nissenbaum has proposed that a discussion of the ethics of data sharing should focus on data *flows* instead of singular data, and take its clue from the informational norms that regulate such data flows in a particular context. Instead of advocating a one-size-fits-all approach of informational privacy, she reinvents the notion of the legitimate expectation of privacy by paying trained attention to what can be legitimately expected within the context(s) in which the data flows take place. More precisely, she suggests distinguishing between norms of appropriateness (what types of data can be shared) and norms of distribution (who gets what information) as two types of informational norms that determine the sharing of information within and between contexts. What makes her framework pivotal for the articulation of informational norms is that she acknowledges that technologies co-constitute existing contexts, and one of the salient points she makes is that new technologies may transform existing contexts and/or create new contexts. This complicates the use of context as a measure for the integrity of information flows, but this complication has the added value of paying homage to the complexity of the Onli*fe* World, instead of reducing the playing field without providing any insight in what is at play.

The crucial 'constituents' of a context where information is shared are defined as: actors (sender, receiver, referent; which may overlap); attributes (types of information; noting that appropriateness of information flows is not one-dimensional,

[36] Vanderstraeten (2007); Hildebrandt (2013b).

[37] Norms and contexts are co-constitutive, Nissenbaum (2010), p. 141.

nor binary);[38] and transmission principles (for instance confidentiality, reciprocity, desert, entitlement, compulsion, need; this entails a rejection of simply dichotomies such as those between access and control). This set of constituents enables developing a privacy impact assessment heuristic (PIA heuristic) that traces the transformation of informational norms due to the introduction of novel technologies, described as socio-technical practices. This heuristic consists of nine steps: (1) describe the new socio-technical practice in terms of information flows (2) identify the prevailing context, (3) identify sender, receiver and referent, (4) identify the principles of transmission, (5) locate applicable entrenched informational norms and identify significant points of departure, (6) make a prima facie assessment, (7) perform the first evaluation in terms of what harms, threats to autonomy, freedom, power structures, justice, fairness, equality, social hierarchy and democracy are expected or have emerged, (8) perform a second evaluation by asking how the system or practices directly impinge on the values, goals, ends of the particular context, (9) dare to formulate a judgment for or against the new socio-technical practice under investigation.[39]

If we refer back to the extended version of the cybernetic definition of informational location privacy, we can check whether the decision heuristics of contextual integrity provides for new insights or a more apt operationalization. The definition was:

> The freedom from networked and/or 'processed' location data of a referent being shared with others without the referent's consent or necessity.

I have inserted the referent that was implied, to make the definition more explicit. Let's be reminded that 'sharing' implies a sender, an addressee, a receiver and an intention. What would it mean to apply the decision heuristic on informational location privacy? I suspect that the relevance of the heuristic will become apparent when testing the negative condition of 'consent or necessity'. Under EU law, consent means 'any freely given specific and informed indication of his wishes by which the data subject signifies his agreement to personal data relating to him being processed',[40] and must be given 'unambiguously' to qualify as a ground for personal data processing,[41] while in the case of sensitive data (revealing racial or ethnic origin, political opinions, religious or philosophical beliefs, trade-union membership, and the processing of data concerning health or sex life) the consent must be explicit to qualify as a valid legal ground.[42] Under EU law, necessity refers to five

[38] Nissenbaum (2010), p. 144.

[39] I believe that in our technology driven world it is becoming increasingly difficult to stand up against technological innovation, cf. Morozov (2013). An unbridled and unsubstantiated technological optimism colonizes our Onli*fe* World. We should, however, dare to accept the responsibility of 'civilizing' the engineers and companies that are reconfiguring our lifeworld. This means that we dare to judge the impact of innovation, after careful scrutiny; it does not—of course—mean that we reject innovation per se.

[40] Art. 2(h) Data Protection Directive D 95/46/EC (DPD).

[41] Art. 7(a) DPD.

[42] Art. 8(a) DPD.

alternative legal grounds: contract, a legal obligation, the vital interests of the data subject, the public interest or the legitimate interests of the so-called data controller (the legal entity that determines the purpose of data processing).[43] The question is, whether the *validity* of the consent or of the various grounds of necessity depends on the context in which data is sent or received. For instance, the answer as to whether consent is an *appropriate* ground for the sharing of 'processed' and/or networked location data, depends on the context. In the context of employment, for instance, I could imagine that the power inequalities between employer and employee render consent inappropriate and therefor invalid. Similarly the business interests of a firm that survives on the sale of inferred location data may not be a proper ground in the context of healthcare or religion.

What makes the decision heuristic of interest here is that it starts with the question of what is new in terms of a socio-technical practice. Rather then trying to develop universal and general rules on the processing of location data, we are asked to first describe the information flows within a new socio-technical practice that implicates location data. If we take the example of Apps on smartphones as a new socio-technical practice, we can describe a series of (new) information flows.[44] These concern location data (temporarily) stored on the device that are sent from the device to app developers, app owners, app stores, Operating Systems and device manufacturer plus third parties such as providers of analytics and advertising networks.[45] It should be clear that we are dealing with observed data, because most users do not intend to send their location data to any of these parties, though they may have provided formal consent in order to get the service they want from the app. This also means that we are talking about messages that are sent from a device to another computing system, enabled by so-called Application Programming Interfaces (APIs) that 'offer access to the multitude of sensors which may be present on smart devices', e.g. 'a gyroscope, digital compass and accelerometer to provide speed and direction of movement; front and rear cameras to acquire video and photographs; and a microphone to record audio. (…) proximity sensors. Smart devices may also connect through a multitude of network interfaces including Wifi, Bluetooth, NFC or Ethernet. Finally, an accurate location can be determined through geolocation services.'[46] Clearly, the different types and the amount of data that is sent indicates that location data can easily be networked with other data (the unique identifiers of the device, content data from the address book, stored pictures, credit card and payment data, phone call logs, browsing history and the more) and further

[43] Art. 7 DPD, sub b-f.

[44] 'Apps are software applications often designed for a specific task and targeted at a particular set of smart devices such as smartphones, tablet computers and internet connected televisions. They organise information in a way suitable for the specific characteristics of the device and they often closely interact with the hardware and operating system features present on the devices.', cf. Art. 29 Working Party Opinion 02/2013 on apps on smart devices, WP202, p. 3.

[45] Art. 29 Working Party Opinion 02/2013 on apps on smart devices, WP202, p. 2, 9–13.

[46] Art. 29 Working Party Opinion 02/2013 on apps on smart devices, WP202, p. 4. Geolocation services have been described in detail in Art. 29 Working Party Opinion 13/2011 on Geolocation service on smart mobile devices, WP185.

processed to enable, for instance, targeted advertising or simply the sale of such 'processed' location data to large data brokers (who may share them with online social networking sites).[47] The relevant information flows are not limited to those between device and app service provider, but will be followed by a number of secondary, tertiary and further flows that are increasingly invisible and unforeseeable (unless in the most abstract way).

We have now described the new practice in terms of information flows (the first step). The second step asks to identify the prevailing context. This means that the answer to the question of whether sharing location data is appropriate cannot be given as a general rule. It depends on the context. If we take the context of travel we can proceed to the next step, taking into account that whatever the heuristic offers will be restricted to the context of travel; to figure out what the heuristic does in another context one has to carefully go through all the steps for that particular context. The third step asks to identify sender receiver and referent. Though we have already discussed that the location data are sent to app developers, OS and device manufacturers, app stores and third parties, we must now pay closer attention to the issue of what agent is doing the sending. Must we pretend that the app user is sending all this networked and/or 'processed' location data, or should we say that the device, the OS, the API or the app itself is the mindless agent? This is an important and interesting question. As far as I am concerned the question is more compelling than the answer. In fact, as mentioned above, we might say that it is the receiver of the data that is 'having the data sent' to itself, thus qualifying as the sender. The app user may in fact be sending location data to her fellow travellers, to her family back home, to potential fellow travellers, or to hotels or other service providers of her choosing. In that case she is obviously the sender of the volunteered location data. This is not so regarding the observed and inferred networked and/or 'processed' location data that is sent to the app developers, the OS or the device manufacturer, the app store or third parties that re-use the data. What is important is to use the third step to, first, investigate whether the messages contain volunteered, observed or inferred location data and what this means for the identification of sender, addressee, receiver and third parties, and, second, to investigate what location messages are sent and/or received by machines, and what location messages are sent and/or received by natural or legal persons. Finally, the point of the exercise is to seek out what new actors enter the context: which senders and/or receivers did not get to send or receive networked or 'processed' location data before the advent of apps on smart devices? It should be clear from the above that in the context of travel a whole series of new actors enters the scene; apart from the fact that people are enabled to send their location to actors they might have shared with even before the advent of smart apps, as we have noted, their networked and 'processed' location data is sent to receivers they may not even be aware of.

The fourth step asks to identify transmission principles. This concerns both the principles that informed the context of travel before the advent of smartphone apps and the emergence of new transmission principles. Instead of falling into the trap of

[47] Hill (2013).

discussing the messages in terms of access or control of location data, the heuristic invites us to check how these apps transform the legitimate expectations of travellers as to confidentiality, reciprocity, desert, entitlement, compulsion and need. Interesting questions arise as to confidentiality: can app users be sure that the location data they send are properly secured against interception? should they understand that their location data are networked and 'processed' by third parties and may be sold to the highest bidder? Reciprocity may come to refer to the fact that app providers make their profits by selling personal data, in return for free services to the referent of those data. This certainly introduces an entirely new kind of reciprocity that is not openly negotiated but entirely implied; there is no clear pricing mechanism that provides transparency as to how the provision of what personal data relates to the service that becomes available. One could of course claim that service providers that render free services 'deserve' to get access to personal data, but this introduces a strange moral connotation into an exchange that has first been commercialized. To what extent are app users entitled to know about what happens to their location data? To what extent are app providers entitled to store and 'process' location data? Which data protection and intellectual property rights conflict at the heart of these novel information flows? Is it still possible to share one's location with others without also providing them to unknown, abstract entities, or are travellers more or less forced to allow the new data flows as a side effect? Can they escape this compulsion by changing the settings of the app, their OS or their device? Is there a need for the multiplication of information flows and it this necessity proportional to the advantages for individual users, also in the long run?

The sixth step involves a prima facie assessment, followed by an evaluation in terms of harms and threats to autonomy, freedom, power structures, justice, fairness and the more. I will not undertake these assessments separately, but will integrate them in the second evaluation that inquires how the sharing of location messages impinges on the values, goals and ends of the context of traveling. This is a tricky business. The context of traveling, obviously, consists of several very different contexts, notably that of business travel, vacation and, for instance, lawful and unlawful immigration (including political and/or economic refugees). The assessment will have to be undertaken in the different sub-contexts, taking into account the values, goals, fairness, power structures and democratic participation that is implied in the case of vacationers, business trips and migration. They may all come to use similar apps and they may all taste some of the less desirable consequences of sharing location data. Customer profiling may cost vacationers money, because companies are enabled to engage in profitable and invisible price discrimination; a business traveller may find that the security of her location data was not guaranteed, allowing competitors to buy networked information they can use against her; a refugee may find himself at the mercy of sophisticated passenger profiling that pre-empt his intention to ask for political asylum.

The point of this exercise is not to attempt a full analysis of the workings of the decision heuristic in the case of networked, 'processed' and 'raw' location data in the context of travel. For such an attempt the voice of those who might be affected would have to be integrated and experts in the relevant context should be involved to explain how the apps may disrupt legitimate expectations. Here, my point was to *show* how the heuristic helps to uncover a plethora of important transformations in our Onl*ife* World that cannot so easily be grasped by applying general rules to individual cases. I believe that this is the crucial distinction between Nissenbaum's contextual integrity as a decision heuristic and the legal framework of data protection within the EU. Whereas the first introduces the concept of context as a constitutive bridge between individual and society, allowing for a more precise exploration of the empirical transformations and their normative implications, the latter remains somehow trapped in the gap between the general rule and the individual case. However, it should also be clear that whereas the decision heuristic provides for numerous occasions for reflection on the ethical implications of data sharing in the Onl*ife* World, it has no teeth, it cannot provide for legal effect; it lacks the conditions that enable the law to provide legal certainty. In the next section I will discuss how the legal concept of purpose binding can help to further interpret informational location privacy, by unravelling the intricacies of the legal principle of purpose binding, as enacted within the EU data protection framework. Before engaging with that, let me conclude by noting that the decision heuristic on contextual integrity has greatly enriched the cybernetic articulation of informational location privacy as 'the freedom from networked and/or 'processed' location data of a referent being shared with others without the referent's consent or necessity'. It has traced the roles and connections of the senders and recipients of 'raw', networked and 'processed' location data, inquired into the transmission principles that 'fit' with a particular context, allowing for a more focused reflection on the difference between sharing volunteered data on the one hand and observed or inferred data on the other. It has thus provided a framework that gives direction to the investigation into the value and the validity of both consent and necessity, depending on the context of application, though taking into account that context has become a moving target. To some extent, the decision heuristic opens a conceptual toolkit to follow the transformations of contexts and their novel interpenetrations. In that sense I believe that it does what the Onl*ife* Initiative aims for: to reengineer our conceptual tools to create greater awareness of the impact of socio-technical change.

## 3.6 The Legal Concept of Purpose Binding

Law, however, is made of different stuff. Though, on the one hand, its procedural justice forces courts to suspend their judgement until the relevant voices have been heard and the facts have been investigated, on the other hand, legal certainty requires a decision. Even when the jury is still out on the ethical standards that should rule individual and institutional actions, courts must give their judgement. As the German legal philosopher Gustav Radbruch noted, people do not necessarily agree on what is morally just and at some point we need a decision that has the force of law, about which standards will orient societal interaction.[48] The law is not only

[48] Radbruch (1950).

after justice, or merely after utility. It also consolidates legitimate mutual expectations between those who may never meet, though they may exchange economic value, share data and contribute in defining the public interest. This signifies one of law's most important dimensions: that of legal certainty, the hallmark of positive law in modern society.[49] It connects the law to the authority of the state, while, in a constitutional democracy, also reigning in its *powers*, which are thus transformed into *competences*: enabling and limiting governments' power to act. This perspective on the law hinges on the intricacies of the internal and external sovereignty of the modern state. The offspring of this dual sovereignty is the so-called *Rechtsstaat* or the Rule of Law, that provides protection of individual liberty and mitigates the monopolistic tendencies of the power of police (in its old meaning of undivided government powers, including administration, legislation and adjudication).[50] We should note, however, that—paradoxically—this protection is dependent on the sovereignty it protects against, and admit that the historical artefact of the Rule of Law cannot be taken for granted in the era of Big Data Space and the Onl*ife* World.

Nevertheless, I will now investigate the notion of purpose binding as a principle that originates in one of the foundational principles of the Rule of Law: the legality principle (not to be confused with its ugly brother, legalism).[51] Legality refers to the fact that governments that are 'under the Rule of Law' can only act on the basis of the law: their legislative, administrative and judicial and other actions must all be based on the law and remain within the limits of the law. Under the Rule of Law the state is both *constituted* by and *limited* by the law. As a consequence, its decisions must be performed for the specific purpose for which a particular competence has been enacted. An explicit and specified purpose thus defines the competence to act, but also—in one and the same Act—restricts governmental actions to those that can be understood to further the relevant goal.[52] The goal is thus both enabling and limiting, in one and the same stroke. The constitutive and the regulative functions of this purpose are two sides of the same coin. Note that legality is not the same as legalism. The latter gives absolute priority to the written code of the legislator, potentially stifling any kind of innovation by requesting adherence to the written Acts of Parliament. The former goes further, by requesting that the legislator itself is under the Rule of Law, requiring that the goals it specifies are legitimate goals—taking into account the written or unwritten constitution and international

[49] Cf. Radbruch (1950), who spoke of the antinomies of the law: legal certainty, justice (as fairness) and the purposiveness or instrumentality of the law.

[50] On the power of police see Dubber and Valverde (2006). With the rise of the modern state in Europe legislation became more important as an instrument to issue general dictates to the subjects of the sovereign. This has been called the rule *by* law. Before the rise of the Rule *of* Law, courts spoke law in the name of the sovereign (*rex lex loquens*); judges were entirely under the rule of man (the king, the Parliament). Only when the courts managed to gain a measure of independence they were capable of standing up against the sovereign, in the name of the sovereign. This is called the paradox of the Rule of Law: *iudex lex loquens*. See e.g. Schönfeld (2008).

[51] On the difference between legality and legalism see (Hildebrandt 2008b), which is my review of Dubber and Valverde (2006).

[52] Cf. e.g. Habermas' *Diskurs-Maxime* which dictates that legitimate actions must be such that they can be reconstructed as being in the general interest (Habermas 1996).

ghts law. This also implies that whenever the state pursues goals in a way atens to interfere with the fundamental rights of individual citizens, such interference must be in accordance with the law, necessary in a democratic society and proportional to the legitimate aim.[53]

It is not clear to me how the principle of purpose binding travelled from constitutional and administrative law to data protection legislation, though it seems an important research question to figure this out. The most important consequence of its migration to data protection is that it becomes applicable to big players that are not (part of) a government. Just like states, legal subjects that process personal data of individual citizens are required to specify a legitimate goal and, just like states, they are accountable for acting within the bandwidth of the purpose they specified. I will now clarify what the principle of purpose binding means in the context of data protection; how it relates to the distinction between volunteered, observed and inferred data; and how it stands with contextual integrity. Finally I will see how both contextual privacy and purpose binding can be framed in terms of sending messages containing 'raw', networked and 'processed' location data.

To understand what the principle means in terms of data protection we must position it in relation to consent, that is often considered to be the foundation of data protection. Within the EU legislative framework, however, the processing of personal data is conditioned by two types of legal requirements:[54] first, there must be a *legal ground* and, second, the processing must be *fair and lawful*. With regard to the first, the data protection directive (DPD) stipulates that one of six legal grounds must apply: only the first concerns (a) freely given and informed *consent*, the other five concern *necessity* in relation to (b) a contract, (c) a legal obligation, (d) the vital interests of the data subject, (e) the public interest or (f) the legitimate interests of the data controller (if these interests are not overruled by the fundamental rights of the data subject).[55] What is important is that *whichever* ground is applicable, the processing of personal data must *always* comply with the conditions of lawful and fair processing, the second type of legal requirements for the processing of personal data. One of these conditions is purpose specification, and another is use limitation, restricting the use of data to what is compatible with the purpose as specified.[56] This means that one *cannot* consent purpose limitation away; a valid new legal ground does not imply that historical data can now be used for an incompatible purpose in relation to the one for which they were originally processed.[57] Purpose binding thus ties whoever processes personal data to the explicit legitimate purpose as it was specified upfront, when the data were first collected. It chains that entity to its own

[53] This is known as the triple test for the justification of interference with the human rights of privacy, freedom of religion and freedom of speech in the European Convention of Human Rights.

[54] Next to a number of other requirements, notably those concerning transparency (information obligations).

[55] Art. 7 D 95/46/EC.

[56] Other conditions see to the integrity of the data, meaning its completeness and correctness. Art. 6 D 95/46/EC.

[57] See on this the Art. 29 Working Group 03/2013 on purpose limitation, WP 203.

stated—and necessarily legitimate—purpose. It should be obvious that this creates a friction with the mantra of Big Data, that seems to require collecting as much data as possible to enable unforeseeable novel correlations that create added value.[58]

The principle of purpose binding is connected with the central role of the data controller, i.e. the legal entity that determines the purpose of the processing of personal data. The data controller is *not* necessarily the entity that actually processes the data; it is, however, responsible for whatever processing is performed under its authority. If we relate this to the idea of a message, we can say that if a user of a location based service (LBS) shares her location data with a restaurant, this user may be termed the data controller, while the LBS is the data processor.[59] This is especially relevant if friends can share the locations of their friends with other friends. However, to the extent that the LBS uses the location data for its own purposes, e.g. for behavioural advertising or any other business model, the LBS is the data controller and is obliged to specify its purposes explicitly, at the latest when it starts processing the data. And, the LBS is not allowed to re-use the data for an incompatible purpose, nor is any other data controller allowed to do this.

All this should clarify that consent is not the most important principle of data protection legislation. Most messages containing 'raw', networked or 'processed' location data are sent for a purpose that is based on necessity: for instance, because a book is bought online the location is sent to enable delivery; or, because employers are legally obligated to send data on travel compensation for their employees to the tax authority; or, because a person is missing in a snow storm and the location of her phone may save her life; or, because a contagious disease requires knowledge of the precise location of contagious people; or, because the business model of a LBS depends on selling 'processed' location data, while it has taken measures to mitigate or even avoid interference with the fundamental rights of its users (for instance by means of anonymisation or pseudonymisation). But besides the fact that most personal data is not shared on the basis of informed consent, *even when it is* it must be processed only for the explicitly specified and legitimate purpose. This implies that to uphold data protection, purpose binding (the combination of purpose specification and use limitation) is the foundational principle, not consent. This is not just the case within the EU jurisdiction. Purpose specification and use limitation are part of the 1980 OECD Fair Information Principles that inspire most data protection regimes on a global level. Perhaps the major exception is the US jurisdiction that makes the application of this principle dependent on sectorial legislation.

Does this mean that in the US the question of whether, how and to what extent purpose binding applies depends on the context? Or should we rather expect that

[58] E.g. Massiello and Whitten (2010) on the added value of function creep (though this is not a term they use to refer to re-using data for novel objectives).

[59] Since it is the LBS that has created and offers the service one can of course argue that it is—for this reason—the 'real' or even the sole data controller. See, however, the Opinion of the Advocate General of the European Court of Justice (EcJ), regarding the question of whether Google, as a search engine, is a data controller or a data processor with regard to the content it indexes and ranks. Cf. the Opinion of Advocate General Jääskinen in Case C-131/12 of 25 June 2013 Google Spain v Agencia Española de Protección de Datos (the judgement of the EcJ is expected in 2014).

even within the EU jurisdiction the content and the scope of the purpose binding principle is largely determined by the context that is at stake? Or, should we understand the purpose binding principle at the global level as a legal instrument to sustain contextual integrity, because it enables data controllers in different contexts to determine different types of purposes? Or should we, finally, determine the scope of the purpose binding principle in view of whether it concerns volunteered, observed or inferred data—independent of context or jurisdiction? As to the latter, one can imagine that in the case of observed data purpose binding is less obvious because the purpose is practically invisible for the data subject, who is hardly aware of all the tracing and tracking that is going on. That might require more stringent application of the principle, but strict application easily irritates data subjects who keep getting messages about whether they agree that their location data is being mined, e.g. to improve the functionality of their navigator.[60] Again, much inferred location data concerns mobility or other patterns at the aggregate level, which means that the legal obligation to comply with the purpose binding principle does not apply, because these patterns do not—by themselves—render an individual identifiable. They rather allow to distinguish, target and discriminate different types of persons, depending on their residence, travel habits, work place, especially when linked with income, spending capacity, religion, sex, education, health. Location data then, is just one data point that helps to infer future behaviours, e.g. earning capacity, health risks or even morbidity.

Instead of providing unilateral answers to the questions I just raised, I will share my intuition that the *legal* obligation to comply with purpose binding has a complex relationship with the *ethics* of contextual integrity. Legally speaking, in the US jurisdiction the applicability of the purpose binding principle depends on the fragmented legal framework of data protection, which seems to differ per context. But this may have little to do with Nissenbaum's decision heuristic. I am not so sure that this heuristic underlies the choices made about whether or not to implement the principle in a particular sector.[61] That being said, the content and the scope of the purpose binding principle will probably vary in different contexts, also within the EU jurisdiction.[62] For instance, in the case of commercial transactions the scope of the purposes that can legitimately be determined by the data controllers (companies) is fundamentally different from the scope in the context of healthcare. The latter requires very precise and narrowly defined purposes to minimize potential

[60] This seems to be the case with regard to the obligation to provide prior informed consent for the use of tracing and tracking mechanisms, as stipulated—since 2009—in the ePrivacy Directive (D 2002/58/EC). On this, art. 29 Working Party, Opinion 02/2013, on providing guidance on obtaining consent for cookies, WP 208; idem, Opinion 04/212 on Cookie Consent Exemption, WP 194.

[61] Though Nissenbaum (2010, pp. 153–156) provides an interesting and convincing example, regarding the regulation of PII in financial transactions.

[62] In deciding whether further processing (re-use) of personal data is still in line with purpose binding requirement, the DPD demands that the purpose of further processing is not incompatible with the explicitly specified purpose for which the data was collected. The decision on whether a new purpose is compatible depends, amongst others, on the context. See, in more detail, Art. 29 Working Party, Opinion 03/2013 on purpose limitation, WP 203.

harm to the mental and physical integrity of patients, even though we should acknowledge that the advent of Big Data Space incentivizes the collection of ever more health-related data and the Onli*fe* World invites people to share health data with their peers in settings similar to SNSs. This relates to the requirement of proportionality between the legitimate aim of data processing and the infringement of e.g. human dignity. The context of commercial transactions seems to allow very broad and vague purpose specifications that include selling personal data for a profit, even though many would object to these practices.

The real problem here seems to be that the context of eCommerce tends to colonize other contexts in the Onli*fe* World, requiring e.g. newspapers, energy saving services and basically any other utility or public interest—including healthcare—to reinvent their 'business case' in terms of the sale of personal data. Purpose binding may blend into this by allowing organisations to reframe their purposes in terms of the added value created by collecting Big Data. To the extent that this happens, purpose binding cannot sustain contextual integrity, precisely because various contexts are overruled by the context of commercial gain.[63]

It is crucial to keep in mind that purpose binding, at least in the EU jurisdiction, is a legal obligation. It is articulated as a side constraint on the processing of personal data. The decision heuristic on contextual integrity, however, is not a legal obligation, but an attempt to frame an ethics of data sharing; it raises a number of empirical as well as normative questions that increase the reflective underpinnings of whatever the outcome is. This enhances the robustness of the outcome. One of the connections between the contextual integrity decision heuristic and purpose binding could therefor be that the rigorous reflection of the first should feed into the interpretation of the second. This should help to prevent a reduction of legality to legalism; when law is separated from ethics it ceases to qualify as law,[64] it becomes administration. Therefor, I believe that the upcoming EU Data Protection Impact Assessment—another legal obligation—could benefit from decision heuristics such as the one developed by Nissenbaum. This should help to inform the *quality* of the purpose specification and the *mindfull compliance* with the subsequent use limitation. Thus, purpose binding also feeds back into the contextual integrity decision heuristic, by means of a careful investigation of what new purposes are enabled by new technologically mediated information flows. Even more to the point, we should investigate how Big Data Space relates to the idea that data controllers can only process personal data for 'old' purposes. As we all know, anonymisation and even pseudonymisation do not resolve this problem, because precisely in Big Data Space increasingly enables deanonymisation.

---

[63] The introduction of the notion of pseudonymous data in the draft General Data Protection Regulation as adopted by the LIBE committee of the European Parliament is highly problematic for this very reason: it assumes that if data controllers have a legitimate interest in the processing of personal data, this processing will be assumed not violate the fundamental rights of the data subjects if the data has been pseudonymised.

[64] This does not imply that law is equivalent with ethics. See Radbruch (2006) on the importance of legal certainty as the distinctive characteristic of law, compared to justice; at the same time, however, Radbruch warns that law that does not strive for justice no longer qualifies as law.

This brings me back to the definition of informational location privacy and the question of its definition in terms of a message. To integrate the principle of purpose binding we can extend the definition:

> Informational location privacy is *the freedom from* 'raw', networked and/or 'processed' location data of the referent, the sender, the addressee or the receiver of a message being shared with others without consent or necessity, and *the freedom from* such location data being shared for purposes incompatible with the explicitly specified and legitimate purpose for which it was first collected.

Please note that this definition is not equivalent with the EU or US legal rights to data protection for location data. It is a definition in terms of the sending of messages, containing 'raw', networked and/or 'processed' location data, not merely about the processing of data. It is about data flows, rather than data processing. It is about 'raw', networked and/or 'processed' location data being send between machines, between humans or between humans and machines; not about intra-machine processing of data. This has advantages and drawbacks, as indicated above. The definition applies to volunteered, observed and inferred location data, but only insofar as they are 'of' the referent, the sender, the addressee or the receiver of the data flows. Insofar as inferred data are patterns, profiles or correlations at an aggregate level, they are outside the scope of the definition, but as soon as they are applied to the referent, sender, addressee or receiver they are part of the definition (under 'processed' location data). The added advantage of this definition, next to the shift from individual data to flows of information, is that it highlights the agency—and patiency—of those involved (sender, addressee, receiver, referent).[65] Instead of treating the agents (sender, receiver) and the patients (referent, addressee) as separate entities, they are viewed here in the context of the relationship they have when sending and receiving messages, and/or when being the addressee or the referent of such messages. The framing of informational location privacy in terms of messages can thus clarify the reciprocity of the relationships, the power structures they involve, the responsibility (liability) for the actions undertaken and the importance of rights for those affected by these messages.

Should we integrate the contextual privacy heuristic into the definition? One answer could be that the heuristic sees to an investigation that should occur before novel technologies are introduced, or while designing legislation to enable and constrain their employment. As developed, the heuristic is not focused on the question of *what is* informational privacy but on *how technologies impact contextual privacy and whether this is acceptable*. As argued above, I think that the notion of contextual privacy can, nevertheless feed into the interpretation of the principle of purpose binding, notably by the courts, by demanding focused attention to what the context of application requires. This would extend the definition as follows:

> Informational location privacy is *the freedom from* 'raw', networked and/or 'processed' location data of the referent, the sender, the addressee or the receiver of a message being

[65] On the salience of thinking in terms of both agency and patiency see e.g. Floridi and Sanders (2004). The agent is whoever acts morally relevant, the patients is whoever is affected in a morally relevant way. The concept of patiency goes back to Aristotle's *Physics*.

> shared with others without consent or necessity, and *the freedom from* such location data being shared for purposes incompatible with the explicitly specified and legitimate purpose for which it was first collected, taking note of what the context *within which* or the contexts *between which* messages are or may be shared requires.

This is too long a definition for everyday usage. It captures what is at stake, but is on the verge of turning into 'legalese'. Nevertheless, I believe that the exercise of developing this definition helps to enrich our understanding of what informational location privacy means. As the reader may know, my own favourite working definition of the right to privacy is 'the freedom from unreasonable constraints on the building of one's identity'.[66] Where the former definition is very detailed, the latter is very abstract. The point is that the sharing of 'raw', networked and/or 'processed' location data in the era of Big Data Space, in an Onli*fe* World, can constrain the building of one's personal identity in numerous ways. We must insist that these constraints are reasonable and one way of determining the reasonableness is to require consent or necessity *as well as purpose binding*, and to investigate how novel constraints fit with the contexts of application.

I conclude with the observation that the biggest challenge to *contextual integrity as a precondition for informational location privacy* may be that the context of commercial benefit and monetary added value seems to colonize all other contexts. If we do not figure out how to preserve the capability of individual citizens to develop legitimate expectations for the sharing of 'raw', networked and 'processed' location data within and between different contexts, the idea of informational privacy may become an empty shell. The threat is not that the institution of context is a moving target, though this presents a formidable challenge. The more complex threat may be situated in the surreptitious colonization of any relevant context by the dictates of commercial enterprise. This 'colonization' can easily turn the PIA decision heuristic as well as the purpose binding principle into lame ducks, e.g. by translating contextual appropriateness or the scope of a compatible purpose into the outcome of a balancing act between anything and economic value—thus nicely complying with the side constraint that stipulates the requirement to perform such a balancing act.

## 3.7 Conclusions: Framing the Balancing Act for Location Messages

One of the objectives of this chapter was an investigation into whether, and if so, under what conditions and how contextual integrity and purpose binding form either side constraints on the free flow of information, or require a balancing act between the civil liberties of individual citizens and the free flow of information. Instead of proceeding straightforwardly to answer this question, I have taken a more oblique way of tackling the issue. It is clear, upfront, that the PIA heuristic and the purpose binding principle are formulated as side constraints: they both require that specific

[66] Cf. e.g. Hildebrandt (2008).

steps are taken and conditions fulfilled before the sharing of 'raw', networked and 'processed' data is either ethically right or lawful. When taking a more in-depth view of both, we encounter various requirements that actually consist of a balancing act, e.g. the last step in the decision heuristic (weighing the impact of novel technologies on the transmission principles regarding 'raw', networked and 'processed' data to decide their acceptibility), and the proportionality test that determines the legitimacy of the personal data processing in relation to the purpose (notably the five legal grounds that involve necessity).

This—prima facie—answer does not bring much news. To generate potentially new perspectives I have formulated the concept of informational location privacy in line with the cybernetic approach to communication, defining it as the freedom from specific types of messages. I have argued that this has three advantages. First, it translates the problem into the language of information theory that is at the root of the development of the computational systems that increasingly determine our lifeworld. This has spurred investigations into Big Data space and the Onl*ife* World. Second, it focuses on information flows instead of discrete data, which is pivotal in the era of Big Data Space and the hyperconnectivity that is prevalent in the Onl*ife* World. Third, it highlights the agency and the patiency of both humans and machines as senders, addressees, receivers and referents of location messages, instead of disentangling these agents/patients from the messages and information flows in which they are implicated. This is again pivotal, in the era of the Onl*ife* World where the heteronomy of human-machine relations transforms both self and society. Taking into account that there are three types of location data: volunteered, observed and inferred, I have extended the cybernetic definition of privacy with an explicit indication of the type of location data at stake. That is, I have distinguished between volunteered and observed location data in itself ('raw' data), volunteered and observed location data that is linked with other data (networked data) and the inferred data resulting from data mining operations on 'raw' and networked location data ('processed' data, which also includes location data inferred from other types of data). I have added the notion of intent, that is implied by the concept of 'sending'. To keep machine-to-machine communication in the loop I view intent at the highest level of abstraction, that includes the mindless intent of machines. Intent implies that next to the sender, the receiver and the referent, the addressee (the intended receiver) becomes part of the definition (which may overlap with the receiver but this need not be the case). Intent also implies that sending a message has a purpose. The fact that a message has a purpose does not necessarily mean that this purpose is achieved, nor does it imply that the receiver has the same purpose as the sender. Precisely by differentiating the actors and patients it becomes possible to stress the role that purposes have in the sending of messages, raising questions about who determines that purpose, who is accountable for its determination and/or compliance and the question of whether and under what conditions the referent of a message can veto messaging if she does not agree with its purpose. The DPD defines all these positions, but it helps to take some distance from the monolithic assumptions that underlie the different roles in the DPD, e.g. by noting that people can be data subjects and data controller with regard to the same location message, depending on who is the addressee (e.g. one's friends) and who is the receiver (e.g. the service provider).[67]

Speaking of location messages is less anonymous than speaking of location information flows, while, like the terminology of location information flows, it spotlights that a location message may be sent from one context but be received in another (whether on purpose, by accident or due to eavesdropping). Thinking in terms of senders, receivers, addressees and referents also helps to understand the importance of the distinction between volunteered, observed and inferred data. Observed data were not sent to the receiver, unless we equate the receiver with the sender (establishing that a data controller that observes behavioural data actually sends that data from the device of the referent to its own servers). Inferred data can be produced by means of intra-machine computations, which does not involve a message, but techniques such as machine learning may infer data by learning from the 'messages' they pick-up from their environment.[68] The most salient form of inter-machine messaging that produces inferred data is that of multi-agent systems, whose emergent behaviour informs so-called simulation games.[69] Such simulations are increasingly employed to design the various layers of automated decision systems, e.g. in the case of the smart grid. The assumptions built into such systems have potentially ground-breaking implications for contextual integrity as they may enable the sending of e.g. 'raw', networked or 'processed' location data outside the context where the initial messages were exchanged. One example is the employment of energy usage data associated with a particular location to detect social security fraud. Another is the use of flexible pricing to incentivize new business models for value added services, which may invite business models that correlate the location of the 'energy user' with lifestyle and purchasing habits.[70]

Purpose limitation thus has its limits. If, at the moment that location data are first observed, volunteered or inferred, a multiplicity of purposes is already specified (e.g. to provide electricity, to send the bill, to detect social security fraud, to send targeted advertising, to enhance customer relationship management with value added services), the protection against a violation of contextual integrity might be nihil. It is, then, a side constraint that can easily be complied with, without provid-

---

[67] The DPD does not exclude this possibility, but the implications are as yet unclear.

[68] This raises the question of whether the feedback that a machine 'receives' has been sent by its environment or is merely 'perceived'. I would suggest that this depends on the measure of agency of whichever part of the environment is either sending or being perceived. On agency at the high level of abstraction of autonomous machines Floridi and Sanders (2004); on human agency as a special type of agency Hildebrandt (2011).

[69] These games may build on traditional assumptions of economic theory, as integrated in game theory, or, alternatively, they may incorporate the insights of behavioural economics and cognitive psychology. Both build on methodological individualism and adhere to the ideal of a rational decision making process. The main difference is that behavioural economics concludes that humans are biased and need help to become rational. The supposed biases described in cognitive psychology are often used to influence—if not manipulate—people's behaviour, notably in the realm of policy science and marketing. Cf. Thaler and Sunstein (2008).

[70] On the implications of profiling in the context of the Smart Grid within the EU jurisdiction Hildebrandt (2013c).

ing substantial protection. From a legalistic perspective this may not be a problem, but such an interpretation of purpose binding flies in the face of the legality principle. Legality means that people have the capability to develop legitimate expectations and this capability is not a character trait but something a society affords by organizing things in one way rather than another.[71] The lack of legitimacy generated by the multiplicity of purposes may be remedied by integrating the decision heuristic of contextual integrity into the decision process on what purposes and what legal grounds are legitimate. Which transmission principles are bent, transformed or eroded to accommodate the plethora of novel purposes that mushroom in the Onli*fe* World? What fairness is tweaked, which reciprocity is broken, does a new purpose unbalance existing power relationships? Or does it increase already existing power asymmetries that cannot be justified? Thus, the substantive protection that purpose limitation aims to provide under the heading of legality may be re-enabled by paying keen attention to the legitimate expectations within and between particular contexts. This solution, however, depends on boundary work between, on the one hand, the contexts of politics, health, employment and others, and that of economic markets on the other hand. In an era where the context of economic markets tends to overrule any other context, we may need to rethink the relationship between partially overlapping spheres of life, otherwise the outcome of any balancing act becomes polluted by the monolithic dictates of one particular logic.

## References


Amoore, Louise. 2011. Data derivatives. On the emergence of a security risk calculus for our times. *Theory, Culture & Society* 28 (6): 24–43 (1 November).

Altman, Irwin. 1975. *The environment and social behavior: Privacy personal space territory crowding*. Montery: Brooks/Cole.

Beresford, Alastair R., and Frank Stajano. 2003. Location privacy in pervasive computing. *Pervasive Computing* 2 (1): 46–55 (January–March).

boyd, danah, and Kate Crawford. 2011. Six provocations for big data. SSRN scholarly paper ID 1926431. Rochester, NY: Social Science Research Network. http://papers.ssrn.com/abstract=1926431. Accessed 9 April 2014.

Cohen, Julie E. 2007. Cyberspace as/and space. *Columbia Law Review* 107:210–256.

De Hert, P., and S. Gutwirth. 2006. Privacy, data protection and law enforcement. Opacity of the individual and transparency of power. In *Privacy and the criminal law*, eds. Erik Claes, Antony Duff and S. Gutwirth. Oxford: Intersentia (pp. 61–104).

Dubber, Markus Dirk, and Mariana Valverde. 2006. *The new police science: The police science in domestic and international governance*. Stanford: Stanford Univ. Press.

Esposito, Elena. 2011. *The future of futures: The time of money in financing and society*. UK: Edward Elgar Pub.

Floridi, Luciano, and J. W. Sanders. 2004. On the morality of artificial agents. *Minds and Machines* 14 (3): 349–379.

Galison, Peter. 1990. Aufbau/Bauhaus: Logical positivism and architectural modernism. *Critical Inquiry* 16 (4): 709–752.


[71] Cf. Robeyns (2005).

Gitelman, Lisa. ed. 2013. *Raw data is an oxymoron*. Cambridge: MIT Press.
Gordon, John. 1984. The Alice and Bob after dinner speech. http://downlode.org/Etext/alicebob.html. Accessed 9 April 2014.
Greenfield, Adam. 2006. *Everyware: The dawning age of ubiquitous computing*. Berkeley: New Riders.
Habermas, Jürgen. 1996. *Between facts and norms: Contributions to a discourse theory of law and democracy (Studies in contemporary German social thought)*. Cambridge: MIT Press.
Hayles, N. Katherine. 1999. *How we became posthuman: Virtual bodies in cybernetics, literature, and informatics*. Chicago: University of Chicago Press.
Hildebrandt, M. 2008a. Profiling and the identity of the European citizen. In *Profiling the European citizen. Cross-discplinary perspectives,* eds. M. Hildebrandt and S Gutwirth. Dordrecht: Springer (pp. 303–343).
Hildebrandt, M. 2008b. Governance, governmentality, police and Justice: A new science of police. *Buffalo Law Review* 56 (2): 557–598.
Hildebrandt, M. 2011. Criminal liability and 'Smart' environments. In *Philosophical foundations of criminal law,* eds. Duff Antony and Green Stuart, 507–532. Oxford: Oxford Univ. Press.
Hildebrandt, M. 2013a. Slaves to Big Data. Or Are We? IDP. Rrvista De Internet, Derecho Y Politica 16 (forthcoming 2014). http://works.bepress.com/mireille_hildebrandt/52.
Hildebrandt, M. 2013b. Profile transparency by design: Re-enabling double contingency. In *Privacy, due process and the computational turn: The philosophy of law meets the philosophy of technology (Hardback)-Routledge*, eds. M Hildebrandt and E. De Vries. Abingdon: Routledge (pp. 221–246).
Hildebrandt, M. 2013c Legal protection by design in the smart grid. Report commissioned by the Smart Energy Collective (SEC). Nijmegen: Radboud University Nijmegen.
Hill, Kashmir. 2013. Facebook joins forces with data brokers to gather more intel about users for ads. *Forbes*. http://www.forbes.com/sites/kashmirhill/2013/02/27/facebook-joins-forces-with-data-brokers-to-gather-more-intel-about-users-for-ads/. Accessed 27 Feb 2013.
Husserl, Edmund. 1970. *The crisis of European sciences and transcendental phenomenology; an introduction to phenomenological philosophy*. Evanston: Northwestern Univ. Press.
Ihde, Don. 1990. *Technology and the lifeworld: From garden to Earth (The Indiana series in the philosophy of technology)*. Bloomington: Indiana Univ. Press.
Ihde, Don. 2008. *Ironic technics*. Denmark: Automatic Press.
Leeuw, Karl Maria Michael de, and Jan Bergstra, eds. 2007. *The history of information security: A comprehensive handbook*. 1st ed. Oxford: Elsevier Science.
Masiello, Betsy, and Alma Whitten. 2010. *Engineering Privacy in an Age of Information Abundance. In AAAI Spring Symposium Series*, 2010, http://www.aaai.org/ocs/index.php/SSS/SSS10/paper/view/1188.
Mayer-Schönberger, Viktor, and Kenneth Cukier. 2013. *Big data: A revolution that will transform how we live, work, and think*. Boston: Houghton Mifflin Harcourt.
Merleau-Ponty, M. 1945. *Phénoménologie de La Perception*. Paris: Gallimard.
Morozov, E., and Evgeny Morozov. 2013. *To save everything, click here: The folly of technological solutionism*. New York: Public Affairs.
Nissenbaum, H. F. 2010. *Privacy in context: Technology, policy, and the integrity of social life*. Stanford: Stanford Law Books.
Prins, Corien. 2006. When personal data, behavior and virtual identities become a commodity: Would a property rights approach matter? *SCRIPT-Ed* 3 (4): 270–303.
Purtova, Nadezhda. 2012. *Property rights in personal data: A European perspective. Alphen aan den Rijn*. The Netherlands: Kluwer Law International.
Radbruch, Gustav. 1950. *Rechtsphilosophie. Herausgegeven von Erik Wolf*. Stuttgart: Koehler.
Radbruch, Gustav. 2006. Five minutes of legal philosophy (1945). *Oxford Journal of Legal Studies* 26 (1): 13–15. http://ojls.oxfordjournals.org/content/26/1/13.short. Accessed 26 Jan 2013.
Ricoeur, Paul. 1976. *Interpretation theory*. Texas: Texas Univ. Press.
Robeyns, Ingrid. 2005. The capability approach: A theoretical survey. *Journal of Human Development* 6 (1): 93–114.

Schönfeld, K.M. 2008. Rex, Lex et Judex: Montesquieu and La Bouche de La Loi Revisted. *European Constitutional Law Review* 4:274–301.
Thaler, Richard H., and Cass R. Sunstein. 2008. *Nudge: Improving decisions about health, wealth, and happiness*. New Haven: Yale University Press.
Vanderstraeten, Raf. 2007. Parsons, Luhmann and the theorem of double contingency. *Journal of Classical Sociology* 2 (1): 77–92.
Varela, F.J., Evan Thompson, and Eleanor Rosch. 1991. *The embodied mind: Cognitive science and human experience*. Cambridge: MIT Press.
Verbeek, Peter-Paul. 2006. Materializing morality. Design ethics and technological mediation. *Science Technology & Human Values* 31 (3): 361–380.
Wiener, Norbert. 1948. *Cybernetics: Or control and communication in the animal and the machine*. Cambridge: MIT Press.
World Economic Forum. 2011. Personal data: The emergence of a new asset class. http://www.weforum.org/issues/rethinking-personal-data. Accessed 9 April 2014.
World Economic Forum. 2012. Rethinking personal data: Strengthening trust. http://www.weforum.org/issues/rethinking-personal-data. Accessed 9 April 2014.

# Chapter 4
# With Great Power Comes Great Responsibility: Proposed Principles of Digital Due Process for ICT Companies

**Dawn Carla Nunziato**

**Abstract** Global Information and Communications Technology companies control an enormous amount of expression on the Internet. More than any individual country, these companies are responsible for making decisions with regard to a vast amount of Internet expression. They have become the de facto sovereigns of cyberspace, with the power to balance freedom of expression against public and private interests, as they make determinations about whether and when to accede to requests to censor speech. Yet, these decision-makers have inadequate guidelines for carrying out these awesome responsibilities that have a dramatic impact on the global contours of freedom of speech. In this chapter, Professor Nunziato argues that ICT companies should adopt and implement a set of procedural guidelines embodying principles of "digital due process" that protect due process rights essential to democratic societies, while respecting the autonomy of each democratic society to determine the contours of substantive free speech rights for its citizens. Professor Nunziato argues that ICT companies should implement a set of five due process principles that are implicit in the free speech and due process jurisprudence of the International Covenant on Civil and Political Rights, the European Convention on Human Rights, and the United States Constitution.

## 4.1 Introduction

Information and Communications Technology (ICT) companies like Google/YouTube, Facebook, Yahoo, and Twitter are in control of an enormous amount of expression on the Internet. More so than any individual country, these companies are responsible for making decisions with regard to a vast amount of Internet expression. They host billions of pages of Internet content, while responding on a daily basis to countless requests from countries and individuals around the world to take down content that is deemed objectionable or illegal. These powerful ICT companies have become the de facto sovereigns of cyberspace, with the power to balance freedom of expression against public and private interests on a day-to-day basis, as they make determinations

D. C. Nunziato (✉)
George Washington University Law School, 2000 H Street, NW, Washington, DC, 20052 USA
e-mail: dnunziato@law.gwu.edu

L. Floridi (ed.), *Protection of Information and the Right to Privacy – A New Equilibrium?*, Law, Governance and Technology Series 17, DOI 10.1007/978-3-319-05720-0_4,

about whether and when to accede to requests to censor speech. In the words of First Amendment scholar Jeffrey Rosen, the "fresh-faced tech executives ... in charge of their companies' content policies... [have] more power over who gets heard around the globe than any politician or bureaucrat—more power, in fact, than any president..."[1] These decision-makers have insufficient guidance and inadequate guidelines for carrying out these awesome responsibilities that have a dramatic impact on the global contours of freedom of speech.

Google, for example, is responsible for facilitating 71 % of the world's Internet searches.[2] As the owner of YouTube, Google is also responsible for hosting 100 h of new video content that users post every minute.[3] This is in addition to its ownership of other widely used applications, including the blog site Blogger, the photo-sharing site Picasa, and the social networking site Orkut. Executives at Google are responsible for making determinations about which controversial content stays up and which comes down. Twitter decision-makers enjoy similar, vast power to determine which of the one billion tweets sent every 5 days get disseminated around the world and which get blocked.[4] The same goes for Facebook, Yahoo, and other global ICT giants.

What guidelines should these companies follow in determining which content to facilitate and which to take down? Under what circumstances should ICT companies accede to governments' or individuals' requests to censor content? How, if at all, should they implement such censorship requests? Given that most of these powerful companies are U.S. based, some have contended that these companies should implement the United States' speech-protective values and refuse censorship requests from other, less speech-protective countries.[5] For example, if France requests that a U.S. based ICT company like Yahoo block content that violates French hate-speech laws, Yahoo arguably should simply ignore the request and export the First Amendment to other countries. But after a French court decision in 2000 rendered Yahoo criminally liable in France for failing to block French citizens' access to certain hate speech content it hosted, Yahoo—and other global ICT companies—began to rethink this approach. Although initially resisting France's power to influence what content Yahoo hosts—resulting in lengthy legal battles—Yahoo ultimately modified its policies to prohibit hosting of hate speech content. Instead of Yahoo exporting the First Amendment, it would seem in this instance that France exported its own less speech-protective laws to the U.S. and the rest of the world. How should global ICT companies respond to countries' requests to censor content in light of the Yahoo-France dispute, in which France asserted the power to seize Yahoo's local assets and detain local Yahoo executives for failing to comply with its laws? Is there a middle ground between imposing the First Amendment on the

[1] Jeffrey Rosen, The Delete Squad: Google, Twitter, Facebook and the New Global Battle Over the Future of Free Speech, April 29, 2013.

[2] http://www.netmarketshare.com/search-engine-market-share.aspx?qprid=4&qpcustomd=0.

[3] http://www.youtube.com/yt/press/statistics.html.

[4] http://www.statisticbrain.com/twitter-statistics/.

[5] See, e.g., Rosen, supra note 1.

rest of the world, and acceding to every other country's speech-restrictive censorship requests? How should global ICT companies balance these myriad competing concerns, amidst vastly conflicting regional free speech regimes?

In this chapter, I argue that ICT companies should adopt and implement a set of procedural guidelines embodying principles of digital due process that protect the due process rights that are essential to democratic societies, while respecting the autonomy of each democratic society to determine the contours of substantive free speech rights for its citizens. Protecting due process rights is the first step in protecting and respecting human rights, which transnational corporations—as well as countries—have a duty to protect. As United Nations' Special Representative of the Secretary-General emphasized in his "Protect, Respect and Remedy" framework, business enterprises as well as nations have a duty to respect human rights.[6] An important part of respecting human rights is respecting the rights of individuals to receiving due process in the protection of their rights. ICT companies should adopt and implement a set of due process principles that, I argue, are implicit in the free speech and due process jurisprudence of the International Covenant on Civil and Political Rights, the European Convention on Human Rights, and the United States Constitution. *Before implementing any country's request that they block content, ICT companies should* (1) *ensure that the requesting country has articulated within its laws a narrow, specific description of what speech is illegal, to confine the discretion of decision-makers and to provide fair notice to individuals of what speech is illegal in the first instance;* (2) *ensure that affected parties have received notice in cases where their speech has been deemed illegal and have had the opportunity to be heard in a fair, independent, and impartial proceeding before a censorship decision is ordered;* (3) *require that the requesting country has issued a narrowly tailored, final judicial decision adjudicating the subject speech as illegal;* (4) *implement the resulting blocking order only within the country mandating such blocking; and* (5) *implement the resulting blocking order in an open and transparent manner.* Below I explore the sources of these principles of digital due process for Internet free speech in the International Covenant, the European Convention, and the U.S. Constitution.

## 4.2 Substantive and Procedural Protections for the Right to Freedom of Expression

While international, European, and U.S. instruments provide for substantive protections for the right to freedom of expression, the language of these instruments allows for substantial discretion to be exercised by different countries in constru-

[6] See Report of the Special Representative of the Secretary-General on the issue of human rights and transnational corporations and other business enterprises, John Ruggie, Guiding Principles on Business and Human Rights: Implementing the United Nations "Protect, Respect and Remedy" Framework. A/HRC/17/31.

ing these provisions. Although there is a resultant wide variation among countries regarding substantive protections for speech—including which categories of speech are protected and which are unprotected—there is a growing convergence regarding procedural protections for speech. Below I first briefly explore the substantive dimensions before turning to the procedural dimensions of protection for freedom of expression.

The International Covenant on Civil and Political Rights (ICCPR),[7] which has been adopted by 167 parties and is considered a binding international law treaty, provides in Article 19 that:

1. Everyone shall have the right to hold opinions without interference.
2. Everyone shall have the right to freedom of expression; this right shall include freedom to seek, receive and impart information and ideas of all kinds, regardless of frontiers, either orally, in writing or in print, in the form of art, or through any other media of his choice.[8]
3. [These rights] may... be subject to certain restrictions, but these shall only be such as are provided by law and are necessary:

    a. For respect of the rights or reputations of others;
    b. For the protection of national security or of public order (ordre public), or of public health or morals.[9]

The right to freedom of expression is also protected under the European Convention on Human Rights, which has been signed by 47 nations. Article 10 of the European Convention provides:

> Everyone has the right to freedom of expression. This right shall include freedom to hold opinions and to receive and impart information and ideas without interference by public authority and regardless of frontiers....
> The exercise of these freedoms... may be subject to such formalities, conditions, restrictions or penalties as are prescribed by law and are necessary in a democratic society, in the interests of national security, territorial integrity or public safety, for the prevention of disorder or crime, for the protection of health or morals, for the protection of the reputation or rights of others, for preventing the disclosure of information received in confidence, or for maintaining the authority and impartiality of the judiciary.[10]

Finally, the First Amendment to the United States Constitution provides that "Congress shall make no law... abridging the freedom of speech, or of the press...." The jurisprudence developed in interpreting and applying these protections for free speech not only has a *substantive* dimension of which categories of speech to pro-

[7] International Covenant on Civil and Political Rights, G.A. Res. 2200, U.N. GAOR, 21st Sess., Supp. No. 16, at 52, U.N.Doc A/6316 (1966) (hereinafter ICCPR).

[8] ICCPR, supra note 7, Article 19.

[9] ICCPR, supra note 7, Article 19. The ICCPR provides further, in Article 20, that any propaganda for war or advocacy of national, racial, or religious hatred that constitutes incitement to discrimination, hostility, or violence, is prohibited by law. Id., Article 20.

[10] *See* European Convention for the Protection of Human Rights and Fundamental Freedoms (ETS 5), 213 U.N.T.S. 222 entered into force Sept. 3, 1953, as amended by Protocol 11 (ETS 155) which entered into force May 11, 1994 (hereinafter European Convention).

tect and which to restrict—which differ from country to country[11]—but also has important *procedural* dimensions, which require that "sensitive tools" be implemented to distinguish between protected and unprotected speech.[12] As free speech theorist Henry Monaghan explains, "procedural guarantees play an equally large role in protecting freedom of speech; indeed, they assume an importance fully as great as the validity of the substantive rule of law to be applied....Whenever [freedom of expression] claims are involved, sensitive procedural devices are necessary."[13]

While there is great variation among countries regarding substantive protections for speech—including which types of speech are illegal—there is more widespread agreement regarding the procedures that are essential to ensure meaningful protections for speech. These procedural protections were recently expounded upon by the U.N. Special Rapporteur on the Promotion and Protection of the Right to Freedom of Opinion and Expression in his May 2011 Report to the Human Rights Council. While recognizing that countries enjoy some discretion to restrict child pornography, hate speech, defamation, incitement to genocide, discrimination, hostility, or violence, the Special Rapporteur explained that:

> Any limitation to the right to freedom of expression must pass the following [multi]-part, cumulative test:
> It must be provided by law, which is clear and accessible to everyone (principles of predictability and transparency);
> It must pursue one of the purposes set out in article 19, paragraph 3, of the [International Covenant on Civil and Political Rights], namely (i) to protect the rights or reputations of others, or (ii) to protect national security or of public order, or of public health or morals (principle of legitimacy);
> It must be proven as necessary and the least restrictive means required to achieve the purported aim (principles of necessity and proportionality)[;]
> [It] must be applied by a body which is independent of any political, commercial, or other unwarranted influences in a manner that is neither arbitrary nor discriminatory, and with adequate safeguards against abuse, including the possibility of challenge and remedy against its abusive application.[14]

Similarly, European and U.S. free speech jurisprudence embodies procedural requirements for any abridgements of the right to freedom of expression. Within the context of United States First Amendment jurisprudence, courts have constructed a powerful "body of procedural law that defines the manner in which they and other bodies must evaluate and resolve First Amendment claims—a First Amendment

[11] *See, e.g.*, Ronald J. Krotoszynski, Jr., *The First Amendment in Cross-Cultural Perspective: A Comparative Legal Analysis of the Freedom of Speech* (2006); Robert A. Sedler, *Freedom of Speech: The United States versus The Rest of the World*, 2006 Mich. St. L. Rev. 377; Stephanie Farrior, *Molding the Matrix: The Historical and Theoretical Foundations of International Law Concerning Hate Speech*, 14 Berk. J. Int'l L 1 (1996).

[12] *Bantam Books v. Sullivan*, 372 U.S. 58 (1963).

[13] *Malinski v. New York*, 324 U.S. 401, 414 (1945) (Frankfurter, J., concurring).

[14] Frank La Rue, *Report of the Special Rapporteur on the Promotion and Protection of the Right to Freedom of Opinion and Expression*, United Nations Human Rights Council, at http://www2.ohchr.org/english/bodies/hrcouncil/docs/17session/A.HRC.17.27_en.pdf.

'due process.'"[15] In so doing, courts have developed "a comprehensive system of procedural safeguards designed to obviate the dangers of a censorship system."[16]

## 4.3 Digital Due Process Principles for ICT Companies

> Digital Due Process Principle 1: Before implementing any country's request that they block content, ICT companies should ensure that the requesting country has articulated within its laws a narrow, specific description of what speech is illegal, to confine the discretion of decision-makers and to provide fair notice to individuals of what speech is illegal in the first instance.

In democratic societies, individuals have a right to conduct their lives so as to conform their conduct to the dictates of the law in order to avoid violations of the law. As the United States Supreme Court has explained, "Living under a rule of law entails various suppositions, one of which is that [all persons] are entitled to be informed as to what the State commands or forbids."[17] This in turn requires laws that clearly and precisely indicate what conduct is illegal, so that individuals can steer clear of such conduct. Because of the paramount importance of freedom of expression to democratic societies, it is especially important that laws regulating speech do so in a narrow and precise manner, to avoid creating a chilling effect on expression. Although democratic countries, as sovereign nations, enjoy the power to determine what categories of speech are illegal (subject to the dictates of the International Covenant and European Convention), they should formulate laws articulating which categories of speech are illegal in as narrow and precise a manner as language permits. Therefore, before implementing a blocking order from a country, ICT companies should require that the requesting country has articulated within its laws a narrow, specific description of what speech is illegal, to confine the discretion of decision-makers and to provide fair notice to individuals of what speech is illegal in the first instance. The International Covenant, the European Convention, and the U.S. First Amendment each provides support for the first proposed principle of digital due process, as I explain below.

The International Covenant requires that any limit on the right to freedom of expression be provided by a law that is transparent, clear, and accessible to everyone, such that it is predictable that one's speech will be subject to regulation. In addition, in its Article 10 jurisprudence, the European Court of Human Rights has also made clear that Article 10 requires that laws restricting speech must be "clear and precise" and must indicate with sufficiently clarity the scope of any legal discretion enjoyed by the decision-maker and the manner of its exercise. Notably, in a 2012 case involving the wholesale blocking of the entire Google Sites platform within Turkey—discussed in greater detail below—the European Court of Human Rights had occasion to reiterate the requirements for laws restriction freedom of

---

[15] *See* Henry Monaghan, "First Amendment Due Process," 83 Harv. L. Rev. 518 (1970).

[16] *See* Monaghan, *supra* note 15.

[17] *Papachristou v. Jacksonville*, 405 U. S. 156, 162 (1972).

expression. The Court explained that individuals whose conduct is affected by laws restricting freedom of expression must be able to foresee the law's consequences, and therefore the law restricting expression must be formulated with sufficient precision to enable any individual to regulate his conduct under the law. This requirement affords individuals legal protection against arbitrary interferences by public authorities with rights guaranteed by the Convention.[18]

In addition, under U.S. First Amendment jurisprudence, laws restricting expression must be articulated in a manner that is clear, precise, and specific. The U.S. Supreme Court has repeatedly held that laws restricting speech that are vague or overbroad are invalid. The Supreme Court has also rejected as unconstitutional any system of censorship that reposits unbounded discretion in the decision-maker to determine whether or not speech is protected.

First, without reference to the substantive categories of which speech can constitutionally be deemed illegal, the U.S. Supreme Court has rejected laws that are framed in vague and imprecise terms—both on First Amendment and on Due Process grounds—because such laws fail to provide clear notice of what speech is prohibited and allow for government officials to exercise standardless discretion. The legislature is required to formulate laws regulating speech "with sufficient definiteness [so that] ordinary people can understand what is prohibited"[19] and "in a manner that does not encourage arbitrary and discriminatory enforcement."[20] The "requirement of clarity in regulation is essential to the protections provided by the Due Process Clause."[21] A law will therefore be struck down as unconstitutionally vague if persons "of common intelligence must necessarily guess at its meaning and differ as to its application."[22]

Laws that do not clearly and precisely define the proscribed content are constitutionally infirm because they are fundamentally unfair. Such laws "trap the innocent by not providing fair warning" of what expression is prohibited and because they impermissibly delegate "basic policy matters to policemen, judges and juries for resolution on an ad hoc and subjective basis, with the attendant dangers of arbitrary and discriminatory application."[23] In particular, the U.S. Supreme Court has explained that vague laws have a chilling effect on expression, as such laws tend to lead citizens to "steer far wider of the unlawful zone than if the boundaries of the forbidden [were] clearly marked."[24] On these grounds, the Supreme Court has, for example, rejected a law that, in part, prohibited "treat[ing] contemptuously the flag of the United States," because it failed "to draw reasonably clear lines between the kinds of… treatment that are criminal and those that are not."[25] Although laws regulating non-expressive conduct may also be struck down on vagueness grounds,

[18] See *Case of Yildirim v. Turkey*, European Court of Human Rights 20 (2012). Pars. 57–59.

[19] *Kolender v. Lawson*, 461 U.S. 352, 357 (1983).

[20] *Id.*

[21] *United States v. Williams*, 553 U. S. 285, 304 (2008).

[22] *Connally v. General Construction Co.*, 269 U.S. 385, 391 (1926).

[23] *Grayned v. Rockford*, 408 U.S. 104, 108–9 (1972).

[24] *Grayned*, 408 U.S. at 109 and n.5.

[25] *Smith v. Goguen*, 415 U.S. 566, 572–73 (1974).

vague laws regulating expression are particularly carefully scrutinized because of the danger of chilling constitutionally protected speech. As the Court has explained, "[b]ecause First Amendment freedoms need breathing space to survive, the government may regulate in the area only with narrow specificity."[26]

The U.S. Supreme Court has also consistently rejected laws that are overbroad—laws that sweep too broadly so as to encompass both unprotected speech and protected speech. For example, a law that criminally prohibited the use of "opprobrious words or abusive language, tending to cause a breach of the peace" was held to be unconstitutionally overbroad, even though it could constitutionally be applied to prohibit certain types of particularly harmful expression, because it could also be unconstitutionally applied to protected expression.[27] In addition, the Supreme Court has invalidated systems for licensing speech that vest unbridled discretion in the initial decision-maker.[28] In *Shuttlesworth v. Birmingham*,[29] for example, the Court evaluated the constitutionality of a parade permitting system that vested the City Commission with the broad discretion to deny parade permits if "in [the Commission's] judgment the public welfare, peace, safety, health, decency, good order, morals or convenience require that [the parade permit] be refused."[30] Because the permitting scheme conferred "virtually unbridled and absolute power" on the Commission, it failed to comport with the essential requirement that any law subjecting the exercise of First Amendment freedoms to a license must embody "narrow, objective, and definite standards."[31]

In summary, the International Covenant, the European Convention, and the United States First Amendment each provides strong support for the proposed principle of digital due process that countries be required to articulate narrow, specific descriptions of what speech is illegal, so as to confine the discretion of the decision-maker and so as to provide fair notice to individuals of what speech is illegal. ICT companies should require that this first principle is complied with before acceding to any country's or individual's request to censor speech.

> Digital Due Process Principle 2: Before implementing any country's request that they block content, ICT companies should ensure that affected parties have received notice in cases where their speech has been deemed illegal and have had the opportunity to be heard in a fair, independent, and impartial proceeding before a censorship decision is ordered.

A fundamental component of living under the rule of law is that an individual be accorded with due process of law before his or her rights or liberties are abridged.

---

[26] *NAACP v. Button*, 371 U.S. 415, 432–33 (1963).

[27] *See Gooding v. Wilson*, 405 U.S. 518 (1972); *Broadrick v. Oklahoma*, 413 U.S. 601 (1973) (to be unconstitutional, overbreadth of statute must not only be real, but substantial as well, in relation to the statute's plainly legitimate sweep).

[28] Such standard-less discretion is an independent ground for finding the law unconstitutional, separate and apart from the absence or presence of a provision for judicial review of the initial decision-making determination, as discussed in greater detail infra.

[29] 394 U.S. 147 (1969).

[30] *Id.* at 149–50.

[31] *Id.* at 150–51.

Due process of law generally requires that an individual be granted the opportunity to state her case before an impartial decision-maker before her rights or liberties are deprived—including her fundamental right to freedom of expression. Accordingly, before an ICT company accedes to a country's request to block content, the ICT company should ensure that the individual has received notice and has had the opportunity to be heard by a fair, independent, and impartial proceeding within that country. The International Covenant, the European Convention, and the U.S. Constitution each provide support for this second proposed principle of digital due process, both in specific provisions protecting freedom of expression and in provisions protecting due process of law and fair trials. Below, I first describe the support these instruments provide for the second proposed principle from their freedom of expression provisions. I then describe the support these instruments provide from their due process and fair trial provisions.

In construing the free speech protections in the International Covenant, the U.N. Special Rapporteur on the Promotion and Protection of the Right to Freedom of Opinion and Expression has explained that:

> Any legislation restricting the right of freedom of expression must be applied… with adequate safeguards against abuse, including the possibility of challenge [before an independent body] and remedy against its abusive application.[32]

The Special Rapporteur has emphasized, in particular, that in order to avoid infringing the freedom of expression rights of Internet users, ICT companies should

> only implement restrictions to [the right to freedom of expression] after judicial intervention …Any determination on what content should be blocked must be undertaken by a competent judicial authority or a body which is independent of any political, commercial, or other unwarranted influences.[33]

Similarly, the European Court of Human Rights in its recent decision in Yildirim v. Turkey has emphasized that any legislation mandating a restriction of freedom of expression in the form of Internet blocking or filtering must embody, at a minimum, a procedure providing for the appeal of such decision to a judicial decision-maker.[34]

Under the U.S. Constitution, it has long has been understood that any government-mandated censorship of speech must be accompanied by judicial review of such a decision in order to be constitutional. Any state-mandated censorship of speech prior to judicial review of the censorship decision constitutes a "prior restraint" that is presumptively unconstitutional. Censorship ordered by the executive branch absence prompt judicial review of such a decision is presumptively invalid. The availability of judicial review of any censorship decision is an essential requirement imposed on any state attempt to censor speech. Judicial review of government

---

[32] Frank La Rue, Report of the Special Rapporteur on the Promotion and Protection of the Right to Freedom of Opinion and Expression, United Nations Human Rights Council, A/HRC/17/27.

[33] Frank La Rue, Report of the Special Rapporteur on the Promotion and Protection of the Right to Freedom of Opinion and Expression, United Nations Human Rights Council, A/HRC/17/27 Par. 47, 70.

[34] *Yildirim v. Turkey*, (Application no. 3111/10) at 27–28 (2012) (citations omitted).

orders restricting speech—and the concomitant notice to affected parties coupled with an opportunity to be heard in a judicial proceeding—is a fundamental procedural requirement within First Amendment jurisprudence. U.S. courts have consistently emphasized the importance of the availability of *prompt judicial review* that affords the affected parties notice and an opportunity to be heard before the judicial decision-making regarding censorship determinations.[35] As the Supreme Court has explained, "because only a judicial determination in an adversary proceeding ensures the necessary sensitivity to freedom of expression, only a procedure requiring a judicial determination suffices to impose a valid final [prior] restraint."[36]

In a number of cases involving government attempts to censor speech, the U.S. Supreme Court has emphasized that a judicial determination is a necessary procedural safeguard in the context of the government abridgement of individuals' freedom of expression. The example of *Bantam Books v. Sullivan*[37] is illustrative. This case involved a censorship system for determining which books were legal and which were illegal—in the absence of a provision for judicial review of such determinations. In that case, the Rhode Island Commission to Encourage Morality in Youth was charged with investigating and recommending prosecution of booksellers for the distribution of printed works that were obscene or indecent. The Commission reviewed books and magazines in circulation, and notified distributors in cases in which a book or magazine had been distributed that the Commission found objectionable and for which removal from distribution was ordered. In reviewing the constitutionality of this scheme, the Supreme Court first explained that "the separation of legitimate from illegitimate speech calls for…sensitive tools" and reiterated its insistence that such censorship schemes must "scrupulously embody the most rigorous procedural safeguards."[38] The Court condemned the fact that, under the scheme at issue, "the publisher or distributor is not even entitled to *notice and hearing* before his publications are listed by the Commission as objectionable [and ordered for removal]," as well as the fact that there was "no provision whatever for *judicial superintendence* before notices issue or even for *judicial review* of the Commission's determinations of objectionableness."[39] The Court concluded that, in the absence of these essential procedural safeguards, the "procedures of the Commission are radically deficient" and unconstitutional.[40]

In sum, the Supreme Court has consistently emphasized the importance of providing affected individuals with notice and an opportunity to be heard in an adversarial judicial proceeding before an individual's right to free speech is abridged. Absent such procedural safeguards, government orders to censor speech are un-

---

[35] *See Thirty-Seven Photographs*, 402 U.S. at 372–74; *Kingsley Books, Inc. v. Brown*, 354 U.S. 436 (1957); *Interstate Circuit, Inc. v. City of Dallas*, 390 U.S. 676 (1968); *Bantam Books, Inc. v. Sullivan*, 372 U.S. 58 (1963).

[36] *See United States v. Pryba*, 502 F.2d 391, 405 (D.C. Cir. 1974).

[37] 372 U.S. 58 (1963).

[38] *Id.* at 66.

[39] *Id.*

[40] *Id.*

constitutional. Under the First Amendment—as well as under the free speech protections accorded under the International Covenant and the European Convention—notice, opportunity to be heard, and a fair determination of one's free speech rights by an independent decision-maker are required.

The European Convention, the International Covenant, and the United States Constitution also contain provisions for due process of the law and fair trials rights that provide individuals fundamental protections before the state can abridge their right to freedom of expression and opinion. Because state-mandated censorship orders result in a deprivation of individuals' right to freedom of expression, fundamental principles of due process and fair trial rights as articulated in the European Convention, the International Covenant, and the U.S. Constitution require that any such deprivation occur only as a result of a fair, independent, and impartial decision-making process in which affected parties are provided with meaningful notice and an opportunity to be heard. These provisions provide further support for the second proposed principle of digital due process—the right of the affected parties to receive notice that their speech has been deemed illegal and to have the opportunity to be heard in a fair, independent, and impartial proceeding before a censorship decision is ordered.

The right to due process of law in general, and to a proper judicial determination of one's rights in particular, is of ancient origin and has its roots in early English and American law. The right to trial by due process of law can be traced to the Magna Carta, which provides that "No freeman shall be ... disseised ... of his liberties... except ... by the law of the land."[41] One of the earliest express provisions for such procedural protections for individual rights is provided in Fifth Amendment to the U.S. Constitution, which provides that "no person shall ... be deprived of... liberty ... without due process of law." Similar language was included in the Fourteenth Amendment to the U.S. Constitution, providing that " No State shall... deprive any person of... liberty ..., without due process of law." Since the late 1800s, procedural due process has been linked to the concept of the rule of law.[42] In the mid-twentieth century, the drafters of the Universal Declaration of Human Rights recognized the importance of protecting due process rights, providing in Article 10 that "Everyone is entitled in full equality to a fair and public hearing by an independent and impartial tribunal, in the determination of his rights and obligations...." The European Convention on Human Rights was the first international human rights instrument to set forth detailed protections for due process and fair trial rights.[43] With respect to the determination of civil rights and obligations, Article 6 of the European Convention provides for the general right to procedural fairness, including a public hearing before a fair, independent, and impartial tribunal that provides a reasoned judgment. Specifically, Article 6(1) states that "In the determination of his civil rights and obligations..., everyone is entitled to a fair and public hearing within a reason-

[41] Richard Clayton QC and Hugh Tomlinson QC, *The Law of Human Rights*, at 708 (quoting clause 39 of the Magna Carta of 1215).

[42] *Id.* at 709.

[43] *Id.* at 706.

able time by an independent and impartial tribunal established by law." Article 14 of the International Covenant similarly provides that "In the determination... his rights and obligations..., everyone shall be entitled to a fair and public hearing by a competent, independent and impartial tribunal established by law." In each of these foundational documents and instruments, procedural due process rights and the right to independent, impartial, and fair judicial determinations of one's civil and human rights are recognized as necessary for the meaningful protection of substantive rights, including the right to freedom of expression. In the words of human rights theorists Richard Clayton and Hugh Tomlinson, "[t]he protection of human rights therefore begins but does not end with fair trial rights."[44]

Article 6 of the European Convention guarantees procedural fairness "whenever there is a 'determination' of a 'civil right or obligation.'"[45] The European Court of Human Rights has repeatedly emphasized the centrality of the rights of procedural due process articulated in Article 6(1) and has affirmed that an expansive view of these rights is fundamental to protecting civil and human rights in democratic societies:

> In a democratic society within the meaning of the Convention the right to a fair administration of justice holds such a prominent place that a restrictive interpretation of Article 6(1) would not correspond to the aim and purpose of that provision.[46]

Similarly, the Due Process Clause of the United States Constitution requires that before an individual's liberty is deprived by the state, she must be afforded certain fundamental procedural protections. The courts have repeatedly held that the protections provided in the Due Process Clause apply specifically in the context of the deprivation of one's First Amendment right to freedom of expression. As constitutional commentators Ronald Rotunda and John Nowak explain,

> In their procedural aspects, the due process clauses require that the government not restrict a specific individual's freedom to exercise a fundamental constitutional right without a process to determine the basis for the restriction... [In particular,] whenever the government seeks to restrain speech, there must be a prompt procedure to determine whether the speech may be limited in conformity with First Amendment principles.

Under the Due Process Clause, before the state deprives an individual of a substantial liberty interest such as the right to freedom of expression, the individual must be accorded at a minimum: adequate notice of the charges or basis for government action; an opportunity to be heard by the decision-maker; a determination by a neutral decision-maker; and a decision based on the record with a statement of reasons for the decision.[47] First, regarding the requirement of notice, when the state is considering impairing an individual's constitutionally cognizable liberty interest—such as

---

[44] *Id. at* 705.

[45] *Id.* at 712.

[46] Delcourt v. Belgium (1970), 1 EHRR 355, para. 25, cited in Clayton and Tomlinson, 824.

[47] See Treatise on Constitutional Law-Substance & Procedure Database updated June 2013, Ronald D. Rotunda, John E. Nowak, Chapter 17. Procedural Due Process—The Requirement of Fair Adjudicative Procedures/II. Deprivations of "Life, Liberty, or Property" for Which Some Process Is Due § 17.4. Liberty, 17.4(c) Fundamental Constitutional Rights.

her right to freedom of expression—notice must be provided: "An elementary and fundamental requirement of due process in any proceeding ... is notice reasonably calculated, under all the circumstances, to appraise interested parties of the pendency of the action and afford them an opportunity to present their objections." Regarding the nature of the determination, "[w]hile different situations may entail different types of procedures, there is always the general requirement that the government process be fair and impartial.... [since] a fair trial in a fair tribunal is a basic requirement of due process."

In summary, the freedom of expression provisions, as well as the due process and fair trial provisions, of international, European, and U.S. instruments provide support for the second proposed principle of digital due process—that before ICT companies implement a country's request that they block content, ICT companies should require that affected parties have received notice that their speech has been deemed illegal and have had the opportunity to be heard in a fair, independent, and impartial proceeding before the censorship decision was ordered.

> Digital Due Process Principle 3: Before implementing any country's request that they block content, ICT companies should require that the requesting country has issued a narrowly tailored, final judicial decision adjudicating the subject speech as illegal.

Before complying with any government's request to censor content, ICT companies should require a court order that is narrowly tailored to achieve the compelling government interest at stake in regulating the speech at issue. Because of the paramount importance of freedom of expression in democratic societies, limitations on this freedom, when imposed, should be imposed as narrowly as possible. Before implementing any country's blocking order, ICT companies should ensure that the order is narrowly tailored. The International Covenant, the European Convention, and the U.S. First Amendment each provide support for this proposed principle of digital due process.

First, as discussed above, the International Covenant's freedom of expression provisions require that laws or orders restricting freedom of expression be narrowly tailored. The U.N. Special Rapporteur on the Promotion and Protection of the Right to Freedom of Opinion and Expression explains that:

> Any limitation to the right to freedom of expression must... be proven as necessary and the least restrictive means required to achieve the purported aim (principles of necessity and proportionality).

In addition, in its First Amendment jurisprudence, the U.S. Supreme Court has repeatedly emphasized that any judicial order regulating speech must be narrowly tailored. Such orders "must be couched in the narrowest terms that will accomplish the pinpointed objective permitted by the [Constitution]"[48] and must "burden no more speech than is necessary"[49] to accomplish their objectives.

Consistent with the importance of freedom of expression to democratic societies, any judicial order restraining expression must be framed in as narrow a manner as

---

[48] *Carroll v. President and Comm'rs of Princess Anne*, 393 U.S. 175, 183 (1968).

[49] *Madsen v. Women's Health Center*, 512 U.S. 753, 765 (1994).

possible to require the surgical blocking of the offending content. Therefore, before implementing any country's request that they block content, ICT companies should require that the requesting country has issued a narrowly tailored judicial decision adjudicating the subject speech as illegal.

Digital Due Process Principle 4: ICT companies should implement a country's blocking order only within the country mandating such blocking.

A foundational principle of national sovereignty is that each nation possesses full control over the affairs within its territorial, geographic boundaries. Under general international law principles, jurisdiction is a nation's assertion of power over the people, properties, and activities within its borders. According to this foundational principle,

> The first and foremost restriction imposed by international law upon a State is that—failing the existence of a permissive rule to the contrary—it may not exercise its power in any form in the territory of another State. [Jurisdiction] cannot be exercised by a State outside its territory except by virtue of a permissive rule derived from international custom or from a convention.[50]

While nations enjoy the power to determine the substantive laws within their own territories, they do not enjoy the right to dictate laws that apply outside of their territories. Thus, any order issued by a court mandating that certain content be blocked should be given effect only within the boundaries of that country.

In recent years, as discussed in greater detail below, certain countries have sought to bring about the worldwide censorship of speech that offended their national laws but that was protected in other countries. Turkey, for example, urged Google to block access throughout the world to content that allegedly insulted the memory of its founder Mustafa Kemal Atatürk—a criminal offense in Turkey. Although Google blocked access to such content for Internet users in Turkey, Turkish officials apparently claimed that this country-specific blocking was insufficient to protect the rights of Turks living abroad. Google properly refused to accede to this additional, overreaching request to export Turkey's laws to the rest of the world. Because countries generally enjoy the sovereign power to dictate the free speech rights of their people only within their borders, Google was right to refuse Turkey's request in this instance. Consistent with the limited sovereignty of each nation, ICT companies should implement a country's valid blocking order only within the country mandating such blocking.

Digital Due Process Principle 5: ICT companies should implement a country's blocking order in an open and transparent manner.

An important part of living in a democratic society governed by the rule of law is that individuals are able to know what the law is and how that law is applied to them. In order to engage in the task of democratic self-government, individuals need to be aware of what the law is and how it is applied, so that they can effectively hold the government accountable for its actions. Governments should adopt,

[50] *The Case of the S.S. "Lotus"*, PCIJ, Ser. A., No. 10 (1927), § 19.

implement, and enforce laws in a public, open, and transparent manner, so that individuals have the meaningful ability to check the power of the government and to hold the government accountable for its decision-making. Laws that are adopted, implemented, or enforced in a secretive or opaque manner violate these principles and thwart the goals of democratic self-government. As de facto sovereigns acting to implement the blocking mandates of countries around the world, ICT companies should implement any such blocking mandates in a manner that is open and transparent, so that affected individuals are made aware that the content at issue is being blocked by the ICT company at the request of their country. Implementing blocking mandates in an open and transparent manner will enable affected individuals to hold their governments accountable and thereby to exercise the rights that are fundamental to individuals living in democratic societies.

The International Covenant, the European Convention, and the First Amendment support the public, open, and transparent implementation of the law and provide support for the fifth proposed principle of digital due process.

The First Amendment provides individuals with the right to access information concerning government decision-making and in particular, with access to judicial records and judicial proceedings.[51] Court decisions and court orders are generally publicly available, so that individuals can hold the government properly accountable for its judicial decision-making. Granting individuals access to information regarding judicial proceedings is essential for citizens to provide an effective check on government power.[52] If the government were to implement judicial decisions in an opaque or secretive manner, this essential component of democratic self-government would be thwarted.

The International Covenant's freedom of expression provisions also provide support for this proposed principle of digital due process. In construing the International Covenant's protections for freedom of expression, the Special Rapporteur has emphasized that in order to avoid infringing the freedom of expression rights of Internet users, Internet intermediaries in implementing any blocking or filtering requests should "be transparent to the user involved about the measures taken, and where applicable to the wider public."[53] Such transparency is necessary to achieve the goals of democratic self-government to enable citizens to hold in check the power of their government.

As the entities responsible for implementing the blocking orders of governments around the world, ICT companies should implement these orders in a manner that

[51] *In re Globe Newspaper Co.*, 729 F.2d 47, 52 (1st Cir. 1984) (establishing a First Amendment right of access to records submitted in connection with criminal proceedings); *Oregon Publishing Co. v. United States Dist. Court*, 920 F.2d 1462 (9th Cir. 1990)(extending qualified right of access to plea agreements and related documents in criminal cases).

[52] *See Globe Newspaper Co. v. Pokaski*, 868 F. 2d 497, 502 (1st Cir. 1989); *In re Globe Newspaper Co.*, 729 F.2d 47, 52 (1st Cir. 1984); *United States v. Antar*, 38F.3d 1348, 1359–60 (3d Cir. 1994).

[53] Frank La Rue, *Report of the Special Rapporteur on the Promotion and Protection of the Right to Freedom of Opinion and Expression*, United Nations Human Rights Council, A/HRC/17/27 Par. 47, 70.

is open and transparent, so that affected individuals can hold their governments properly accountable for their actions.

## 4.4 ICT Companies' Current Policies Regarding Responses to Countries' Blocking Orders

Global ICT companies exercise great discretion in determining whether to accede to content removal orders issued from countries or individuals around the world. Although companies like Google, Microsoft, Yahoo, Facebook, and Twitter provide a wealth of information regarding the requests they receive and their actions in response to these requests, they tend to provide far less information about the guidelines they follow in determining whether to comply with such requests. While Google for example, is to be commended for publishing its semiannual Transparency Reports with detailed information about requests it has received from governments to remove content and Google's actions in response to such requests,[54] the only information to be gleaned from its transparency reports is that Google may choose to remove content at the request of a government body when such content violates local law, but it may also choose not to remove such content.[55] YouTube reports that it will accede to governments' request to remove content if and only if the content is in violation of YouTube's Community Guidelines, which prohibit a variety of categories of speech (including categories of speech that are protected by the First Amendment, like "insulting generalizations about people of a particular nationality".)[56] Facebook provides users with a list of rights and responsibilities and indicates that it has the discretion to remove content that is in violation of these rights and responsibilities.[57] Yet, Facebook has been relatively opaque in its determinations of what content to remove.

By comparison, Twitter has been both transparent and clear regarding the policies it has adopted governing content removal. In January 2012, Twitter announced

[54] http://www.google.com/transparencyreport/removals/government/.

[55] See for example Google's Answer to the Frequently Asked Question: " Q. Why haven't you complied with all of the content removal requests? A. There are many reasons we may not have removed content in response to a request. Some requests may not be specific enough for us to know what the government wanted us to remove (for example, no URL is listed in the request), and others involve allegations of defamation through informal letters from government agencies, rather than court orders. We generally rely on courts to decide if a statement is defamatory according to local law."

[56] See http://www.youtube.com/t/community_guidelines.

[57] See https://www.facebook.com/legal/terms (providing, inter alia: "You will not bully, intimidate, or harass any user. ...You will not post content that: is hate speech, threatening, or pornographic; incites violence; or contains nudity or graphic or gratuitous violence.......You will not use Facebook to do anything unlawful, misleading, malicious, or discriminatory....You will not post content or take any action on Facebook that infringes or violates someone else's rights or otherwise violates the law....We can remove any content or information you post on Facebook if we believe that it violates this Statement or our policies.").

its adoption of a relatively speech-protective policy by which it will censor speech within the country requesting censorship upon receipt of a valid order from the country mandating censorship. Under the policy, Twitter will provide notice to affected parties (both content providers and would-be recipients of such content) of the censorship (unless it is prohibited from doing so by law). Twitter sets forth its content removal policy as follows:

> Many countries… have laws that may apply to Tweets and/or Twitter account content. [I]f we receive a valid and properly scoped request from an authorized entity, it may be necessary to reactively withhold access to certain content in a particular country from time to time.
> … Upon receipt of requests to withhold content, we will promptly notify affected users unless we believe we are legally prohibited from doing so (for example, if we receive an order under seal). We also clearly indicate within the product when content has been withheld. And, we have expanded our partnership with *Chilling Effects* to publish … requests to withhold content—unless, similar to our practice of notifying users, we are legally prohibited from doing so.
> Withheld Tweets:
> If you see a grayed-out Tweet in your timeline… or on another user's account…, it means that access to that Tweet has been withheld in your country.
> Withheld accounts:
> Similarly, if you see a grayed-out user in your timeline… or elsewhere on Twitter…, access to that particular account has been withheld in your country.
> … Upon receipt of a request to withhold content, Twitter will attempt to notify affected users of the request via the email address we have on file, identifying the specific content withheld and the origin of the request, in addition to marking withheld Tweets and/or accounts with a visual indicator.

Twitter's content removal policy has many virtues. It is speech-protective in that Twitter will only withhold access for individuals within the country making that request, and it will provide notice both to the Twitter account holder and the would-be recipients of the content that the content has been withheld and the origin of the request for withholding the content. The policy therefore complies with the fourth and fifth proposed principles of digital due process, requiring the surgical implementation of blocking orders by the ICT company only within the country mandating such blocking; and the implementation of such blocking order by the ICT company in a manner that is open and transparent. However, the provision of the policy in which Twitter responds to "valid and properly scoped request from an authorized entity" leaves some room for interpretation. Depending on how Twitter interprets its requirement of a "properly scoped request," this requirement may be consistent with the first and third proposed principle of digital due process, requiring that countries articulate narrow, specific descriptions of what speech is illegal, and requiring a narrowly tailored, reasoned final judicial decision adjudicating the subject speech as illegal. Twitter's content removal policy is deficient in that it merely requires a valid and properly scoped request from an *authorized entity*. It apparently does not require a fair, independent, and impartial *judicial determination* of whether the content is illegal within the country requesting that the content be withheld. Indeed, as discussed below, in its first action to withhold content under this policy, Twitter agreed to withhold content upon request from a non-judicial authority in Germany, as discussed below. This aspect of Twitter's policy does not comport with the second proposed principle of digital due process—ensuring that affected parties have the opportunity to be heard in a fair, independent, and impartial judicial proceeding before a censorship decision is ordered—and is insufficiently protective of free speech, as discussed in greater detail below.

## 4.5 Case Studies and Recommendations for Implementation of Principles of Digital Due Process

Having articulated a set of proposed principles of digital due process that ICT companies should adhere to in responding to censorship requests from governments around the world, I now examine how the implementation of these principles would have proceeded in the context of several recent cases involving restrictions on Internet speech.

### *4.5.1 Case Study 1: Yildirim v. Turkey*

In *Yildirim v. Turkey*, Ahmet Yildirim, a national of Turkey and doctoral student, sued the Republic of Turkey for violating his free speech rights under Article 10 and his fair trial rights under Article 6 of European Convention. The difficulties arose for Yildirim in June 2009 when another website—with whom Yildirim had no connection—posted content via Google Sites (a website creation platform) that allegedly insulted the memory of Mustafa Kemal Atatürk (founder of the Republic of Turkey), which constitutes a crime in Turkey. In response, the Denizli Criminal Court of the First Instance, pursuant to its law regulating Internet publications, ordered the blocking of the offending website, as a preventive measure in the context of criminal proceedings against the site's owner. The Court then sent its order requiring the blocking of the offending website to the Telecommunications and Information Technology Directorate (the "TIB") for execution. The TIB, upon receiving this narrowly targeted blocking order, complained that it did not have the means of only blocking the offending site, and instead requested that the Court modify its mandate to order the blocking of all access to Google Sites in its entirety. The Court complied, modifying its order to require the blocking of Google Sites in its entirely. The TIB then implemented this order and rendered all Google Sites content inaccessible within the country. At no time in this process did the Court or the TIB notify Google or request that Google Sites render the offending site inaccessible within Turkey. Once the TIB rendered the entire Google Sites platform and all of its content inaccessible, Yildirim—who used Google Sites as a platform to publish his academic work—was unable to access his content, including his academic articles and commentary, on his Google Sites website (available at http://sites.google.com/a/ahmetyildirim.com.tr/academic/). Yildirim applied to have the Court's blocking order modified and narrowed, in favor of a method of implementation that would make only the offending website inaccessible, such as by blocking the offending site's URL. The Court dismissed Yildirim's application for a modified order, explaining that the TIB had insisted that the only means of blocking access to the offending website was to block access to the entirety of Google Sites. Three years later, after the criminal case against the owner of the offending website was dropped, the entirety of Google Sites remained blocked within Turkey and Yildirim's Google Sites website remained inaccessible within Turkey. Seeking redress, Yildirim brought an action against the Republic of Turkey, alleging violations, inter alia, of Article 10 and 6 of the European Convention.

How might this situation have been resolved in a less speech-unfriendly manner? Assuming that countries have the right to block within their country—or have blocked within their country—web content that is illegal within their country, first, courts should craft blocking orders in as narrow and specific a manner as possible, and each global Internet platform should implement such judicial orders in as narrow and specific a manner as possible. In the Yildirim case, the Turkish criminal court—after determining that the content on the offending site violated its law prohibiting clear, precise, and narrowly drawn categories of content, and after giving the owner of the offending site notice of the charge and an opportunity to be heard by the court—should have ordered the TIB to block access within Turkey to the offending site—and only the offending site. If the TIB was unable surgically to block access only to the offending site (as the TIB claimed was in fact the case), the TIB or the court could have requested that Google block access only to the offending website and only within Turkey.[58]

Once presented with a request from the Turkish criminal court or the TIB in furtherance of the court's order, Google could have taken steps to ensure that the first

[58] In the past, the Turkish government has expressed dissatisfaction with country-specific blocking, claiming that Internet content illegal within Turkey—like the content at issue in the offending website that allegedly insulted the memory of Atatürk—should be rendered inaccessible by Google for all Internet users, including those outside of Turkey. For example, in June 2008, after Google agreed to block access within Turkey to a series of videos on YouTube that a Turkish court held were violative of Turkish law, see http://www.bbc.co.uk/news/10480877, the Turkish government was apparently unsatisfied and demanded that Google implement worldwide blocking of such content, claiming that a worldwide block was necessary to protect the rights and sensitivities of Turks living outside of Turkey. See Jeffrey Rosen, Google's Gatekeepers, The New York Times, November 30, 2008. When Google refused to expand the scope of the geographic scope of the block to all countries, the Turkish government decided to block access to all of YouTube, which it proceeded to do for the next two and a half years, until the offending videos were removed from YouTube (by a party other than YouTube itself). See http://www.bbc.co.uk/news/technology-11659816. Indeed, Turkey blocked not only YouTube but also a host of other Google services, because it was apparently unclear which of Google's designated IP addresses it was using for YouTube services and which it was using for other services. See http://www.bbc.co.uk/news/10480877. In such a case, it is unreasonable for Turkey to seek to export its country-specific laws regarding illegal content, and it is further unreasonable to block access to the entire video-sharing platform of YouTube—and other unrelated Google services—because of a small handful of offending websites on YouTube that YouTube had already agreed to block for residents of Turkey.

four digital due process requirements were met: first, that the applicable country's law articulates a narrow, specific description of what speech is illegal; second, that affected Internet users received meaningful notice of categories of illegal speech and had the opportunity to be heard by a court *before* their right to freedom of speech/information is abridged; third, that decisions to block content were made in an impartial, independent judicial proceeding; and fourth, that the court's decision was narrowly tailored to avoid collateral censorship/overbreadth and specified precisely which Internet speech is illegal. Once it was satisfied that the country and its courts met these first four requirements of digital due process, Google should then implement its technology in general and the blocking order in particular in a manner that comports with the fifth principle of digital due process by first, implementing the court's decision in a manner that is narrow, open and transparent, so as give meaningful notice to those seeking access to the blocked site that the site was blocked in accordance with a valid court order.

### *4.5.2 Case Study 2: Twitter and Blocking of Neo-Nazi Tweets*

In October 2012, Twitter received from German police a request to close the account of the neo-Nazi organization of Besseres Hannover. The German police informed Twitter that Besseres Hannover "is disbanded, its assets are seized and all its accounts in social networks have to be closed immediately." The police asked that Twitter block Besseres Hannover's account and prevent it from opening alternate accounts.[59] In the first instance of its implementation of its Country-withheld content policy, Twitter responding by blocking the tweets of the organization within Germany, but declining to close the organization's account. Twitter also provided notice to German users that this organization's tweets were blocked within Germany, and provided access to the German authorities' documents requesting the block. Under the policy, Besseres Hannover's tweets appear to German users as greyed-out boxes with the words "@Username withheld" and "This account has

[59] See https://www.chillingeffects.org/notice.cgi?sID=625342 (providing letter from the Head of the Police Administration Department, Hannover, Germany, to Twitter, which indicates: "the Ministry of the Interior of the State of Lower-Saxony in Germany has banned the organisation "Besseres Hannover". It is disbanded, its assets are seized and all its accounts in social networks have to be closed immediately. The Public Prosecutor (State Attorney's Office) has launched an investigation on suspicion of forming a criminal association. It is the task of the Polizeidirektion Hannover (Hannover Police) to enforce the ban. The organisation "Besseres Hannover" uses the Twitter account besseres-hannover@hannoverticker https://twitter.com/hannoverticker. I ask you to close this account immediately and not to open any substitute accounts for the organisation "Besseres Hannover".)

been withheld in: Germany."[60] Twitter has also received requests from countries to make available the identities of individuals who post illegal content.[61]

How might this situation have been resolved in a less speech-unfriendly manner? Although Twitter's implementation of its "country-withheld content" policy is relatively speech-protective, the policy suffers in that it does not require that the content's illegality be determined in the context of a fair, impartial, and independent judicial determination. Twitter should only block access to speech within a particular country upon the receipt of an order that results from an impartial, independent judicial proceeding determining that speech was illegal in that country.

### *4.5.3 Case Study 3: Turkey's Demand that YouTube Block Access Throughout the World to Video Content that Violated Turkish Law*

The Republic of Turkey has had a tumultuous relationship with Google/YouTube. In March 2007, a Turkish judge ordered the nation's telecommunications providers to block access to all of YouTube in response to videos that allegedly insulted the founder of the Republic of Turkey Mustafa Kemal Atatürk, which is a crime in Turkey. The video that initially sparked the controversy was a parody news broadcast that declared, "Today's news: Kamal Atatürk was gay!'" posted by Greek soccer fans to insult their Turkish rivals. The ban on YouTube was ordered and implemented by Turkish officials without consulting with Google/YouTube and without asking the company to surgically block access to only the offending video. The offending video was eventually taken down, apparently by the individuals who initially posted it, but even after it was taken down, Turkish prosecutors found dozens of other YouTube videos that they claimed insulted either Atatürk or "Turkishness," in order to justify the continued blocking of YouTube in its entirety.

Upon learning that access to all of YouTube was being blocked in Turkey, Google executives worked to develop a solution that would placate Turkish officials. The executives set about attempting to determine which videos were clearly in violation of Turkish law prohibiting insults to Atatürk or Turkishness, and then blocking access to those videos within Turkey. Google's plan seemed to be satisfactory to Turkish authorities for a period of time, but then in June 2007, a Turkish prosecutor demanded that Google block access to the offending videos not just in Turkey but throughout the world, asserting the need to protect the rights and sensitivities of Turks living outside the country. Google refused to implement this extraterritorial

[60] http://www.slate.com/blogs/future_tense/2012/10/18/twitter_censors_neo_nazi_group_besseres_hannover_is_first_user_blocked_under.html.

[61] In January 2013, a French court ordered Twitter to identify people who had posted anti-Semitic and racist entries on the social network. See http://www.nytimes.com/2013/01/25/technology/twitter-ordered-to-help-reveal-sources-of-anti-semitic-posts.html?_r=0.

blocking mandate, and in response, the Turkish government once again blocked access to YouTube in its entirety.

How might this situation have been resolved in a less speech-unfriendly manner? First, Google should determine whether Turkey has articulated its laws prohibiting attacks on Atatürk and Turkishness in a narrow and precise manner, so as to provide individuals within Turkey with adequate notice as to what content is prohibited. Second, Turkish courts should provide the owners of offending sites with notice of the charge and an opportunity to be heard by the court determining whether their content violated Turkish law. Third, if warranted, the Turkish court should craft any blocking orders in as narrow and specific a manner as possible with the least impact on freedom of expression. Google/YouTube should not block access to content on its own, without a valid court order identifying specifically which content is to be blocked. Fourth, Google should implement any valid court blocking orders only with respect to individuals within Turkey, rendering such content inaccessible only for individuals within Turkey. The Turkish government's request to block access to such sites throughout the world—in order to protect the rights and sensitivities of Turks living outside the country—oversteps that sovereign's authority and jurisdiction. Fifth, Google should implement any valid court blocking order in a manner that is open and transparent and provides notice to affected individuals that the requested content has been blocked because of a court order, so that those individuals can hold their government properly accountable for its speech-restrictive decisions.

## 4.6 Conclusion

Global ICT companies like Google, Facebook, Twitter, and Yahoo enjoy great power to control what expression is facilitated and what expression is censored in the global marketplace of ideas. And, in the words of Voltaire (and Spiderman), with great power comes great responsibility. ICT companies should wield their great power responsibly, which means adhering to the principles of digital due process that are implicit in the foundational instruments of the International Covenant of Civil and Political Rights, the European Convention on Human Rights, and the United States Constitution.

# Chapter 5
# The Political Economy of Data: EU Privacy Regulation and the International Redistribution of Its Costs

**Hosuk Lee-Makiyama**

**Abstract** In the wide-ranging policy debate on data privacy, the economic impact of regulation has thus far received very little attention. Yet data privacy legislation has the potential to affect different groups and countries asymmetrically, leading to important redistributive effects. This paper aims to illustrate the economic impact of data regulation, using the EU's General Data Protection Regulation (GDPR) as an example, and using econometric methods (GTAP) commonly used in trade economics. GDPR introduces restrictions on cross-border trade flows, which affects input prices to the service industry, and in turn, EU exports, with a direct welfare effects on the EU is a loss of up to € 260 per European citizen. The findings of this study have important implications for the discussions around policy design and regulatory efficiency. The severe economic impact of the GDPR proposal demonstrates that data connectivity and economic interdependency effectively limit the policy space for non-economic regulation. Moreover, this paper shows that mercantilist data restrictions are counterproductive and affect the protecting market more than those who are restricted in accessing it.

## 5.1 Introduction[1]

The internet is arguably one of the largest contributors to the last two decades' economic growth. Analysis in international trade sometimes make the claim that the internet and the free flow of data it has enabled mark the biggest advancement in international trade facilitation since air travel—[2] and indeed, cross-border data processing and data flows are now critical to most commercial activities in both e-commerce and traditional brick-and-mortar business, from whom the internet

[1] Assisted by Bert Verschelde.

[2] See also Lee-Makiyama (2013).

H. Lee-Makiyama (✉)
European Centre for International Political Economy (ECIPE),
Rue Belliard 4–6, 1040 Brussels, Belgium
e-mail: hosuk.lee-makiyama@ecipe.org

L. Floridi (ed.), *Protection of Information and the Right to Privacy – A New Equilibrium?*,
Law, Governance and Technology Series 17, DOI 10.1007/978-3-319-05720-0_5,

serves as both a marketplace and a supply chain for innovation, production and distribution. However, remarkably little of the policy debate concerning internet regulations takes into account the economic impact from these regulation, and data privacy regulations are not an exception: in Europe, the debate has its remits in the open-ended question of a relatively new, uniquely European (supra-national, yet sub-universal) fundamental right that encompasses privacy.[3] This is further complicated by the government deployment of large-scale electronic surveillance, another by-product of the data-driven society, and the degree of complicity by foreign-owned commercial actors has led to fears about the exposure to data collection conducted by foreign-invested firms.

This article discusses data privacy from a purely economic perspective, using the EU General Data Protection Regulation (GDPR) of January 2012 as an example. It is not a misplaced attempt to quantify the monetary value of fundamental rights or national security—however, data privacy legislation does have an economic impact, and that impact can affect different groups and countries asymmetrically, leading to a financial redistribution between countries, industry sectors, producers and consumers. In such cases, the disciplines of law and economics postulate that if all other conditions are alike, legislation with lesser total societal costs must be a more efficient legislation. Indeed, laws are designed to minimise negative externalities or costs, but given the global nature of the legislative subject—the internet—these costs could appear across borders. In this global context, Europe is a unique subject for analysis. Despite the recent economic contraction and the relative decline of Europe, the EU (EU) remains the world's largest economy by a considerable margin in economic output and exports. Any changes to Europe's interaction with other economies in an interdependent system will not only affect the EU itself, but every other country within the international trading system.

This paper uses the methodology of applied trade economics to assess this impact by estimating the economic effects of new regulatory costs and trade barriers. Such barriers lead to changes in prices on goods, services and input factors, which consequently alter a country's economic competitiveness and productivity. This leads in turn to changes in a country's imports and exports, resulting in a recursive process which affects the wealth of nations. The second part of the analysis discusses the redistribution primarily between the EU and the rest of the world.

## 5.2 Data Privacy as Market Regulation

The extent to which personal data is disseminated in a modern economy warrants a long-overdue ethical and legal overview: the current EU directive on data protection dates from 1995,[4] long before many of today's social processes were redefined

[3] Charter of Fundamental Rights of the European Union, 2000/C 364/01.

[4] On The Protection of Individuals with Regard to the Processing of Personal Data and on the Free Movement of Such Data, Directive 95/46/EC.

by the internet. It also moved the primacy from entities (the subject of most regulations) to interactions between entities,[5] while these interactions are increasingly commercial and transnational in their nature.[6] EU governance of data flows is also primarily a commercial interaction. The EU is a free-trade customs union in its origins and a market supervisory body in its construct—but the data privacy directive that has been in force since 1995 even explicitly uses the term 'free movement' of data—implying a commercial freedom that functionally follows the freedom of establishment,[7] movement of services,[8] or the notion of non-discriminatory market access under its obligations to the World Trade Organization.[9]

Privacy rules may indeed have a macroeconomic impact by influencing the decisions of businesses and consumers, and thereby affecting market behaviour. This impact is determined by a number of factors. Firstly, it is affected by the data dependency of an economy. For example, a purchase made in a shop requires card processing, banking and other financial services that provide personal data to the retail point. The shop may also depend on leasing, shipping, utility and facility management services that all use information about the shop's proprietor may used for credit ratings, or other analytical purposes. The goods that were sold to the customer in the transaction were most likely manufactured using electricity and other utilities. Furthermore, various consulting, engineering or creative services may have been involved in a process, drawing on customer data for design, customisation, or to provide guarantees or returns.

In this manner, data processing has become one of the most important 'raw materials' for the services industry, accounting for 15–58 % of the input value depending on the sector.[10] Should the regulatory and administrative burden on suppliers be increased or the presence of foreign competition limited by the cost of data processing, the cost increases will be reflected in prices in the services sectors (which includes retailing, logistics, transport, research, construction, telecommunications, banking, insurance and business or professional services) that account for over 70 % of all economic activities in the EU when measured in GDP.

In turn, these highly data-dependent services sectors account for 15–30 % of the inputs in European manufacturing,[11] making services the most important 'raw material' of the manufacturing process. Furthermore, much European manufacturing (especially the competitive parts of the economy) is highly dependent on exports and foreign markets—the EU remains a dominant trading entity as world's largest

[5] Floridi (2013).

[6] Inter alia, OECD (2005).

[7] Article 49, Treaty on the Functioning of the European Union (TFEU).

[8] Article 56, TFEU.

[9] WTO General Agreement on Trade in Services (GATS); note the exception for 'the protection of the privacy of individuals in relation to the processing and dissemination of personal data and the protection of confidentiality of individual records and accounts' under Article 14.2. For its interpretation, see Hindley, Lee-Makiyama, Protectionism Online, ECIPE Working Paper, 12/2009.

[10] GTAP8, 2007.

[11] World Input Output Database, 2013.

exporter of both goods and services. EU exports represent 45 % of its GDP, and it is evident that a regulatory change in one of its underlying production factors—whether it is cost of electricity, data or labour—changes the competitiveness and trading patterns for not only the EU, but also every other country with which it trades or competes.

The importance of the internet and services is consistent with non-academic research that suggests that the internet has contributed more than one-fifth of all GDP growth in recent years.[12] In comparison, in China, where services make up just 43 % of the GDP,[13] the internet accounted for a mere 3 % of economic growth.[14] Furthermore, the link between data and trade is not a novel concept: UNCTAD, the UN trade body, has estimated that more than half of services exports are enabled and dependent on the existence of information and communications technology (ICT).[15]

## 5.3 The Law and Economics of Data Regulation

The European Commission states assess GDPR will have positive effects on the economy as 'strong data protection can build consumer confidence and strengthen the potential of the market'.[16] The actual level of increased consumer confidence and its translation to increased consumption and other economic benefits hinges on consumer psychology, which is difficult to establish scientifically *ex ante*. It is, however, possible—albeit with broad margins of error—to establish the costs of the regulation to society. Like any legislation, GDPR also introduces additional obligations, uncertainties and liabilities on EU-based businesses. This part of the analysis is based on impact assessments by EU Member States and the European Commission, and does not contradict their conclusion. However, the cost accounting methodology employed in these assessments do not take into account the dynamic effects of squeezing foreign competitors out of the EU, which leads to higher market prices inside the EU for data processing. The relative competitiveness of the EU can be spurred by the legislation, if these benefits exceed the direct and dynamic costs combined.

These effects from GDPR were estimated by Bauer et al. (2013),[17] by using GTAP8—a computable general equilibrium model frequently used for international

[12] McKinsey Global Institute (2011).

[13] World Bank (2013).

[14] See note 12.

[15] UNCTAD (2009).

[16] European Commission, Impact Assessment, Regulation of the European Parliament and of the Council on the protection of individuals with regard to the processing of personal data and on the free movement of such data (General Data Protection Regulation) and Directive of the European Parliament and of the Council on the protection of individuals with regard to the processing of personal data by competent authorities for the purposes of prevention, investigation, detection or prosecution of criminal offences or the execution of criminal penalties, and the free movement of such data, SEC(2012) 72 final.

[17] Bauer et al. (2013).

economics and trade analysis.[18] In the model, any regulatory differences work as a trade cost—a non-tariff barrier—in trading with the EU. It also hampers foreign investments and reduces foreign competition in the EU, leading to price increases. The model applied the administrative burden of GDPR to the services sectors only, and according to the extent data processing is used to produce that service. Manufacturing and agriculture sectors are affected only indirectly through higher cost of production from price increases in services they use.

In this paper, two scenarios are used, based on the results of the study by Bauer et al. The first, lower-bound scenario factors in the extra administrative burden of the new regulation for EU businesses based on public impact assessments.[19] Such costs include the obligation to employ data protection officers in businesses; to carry out data protection impact assessments; to notify the supervisory authority of all personal data breaches; and the administrative cost of demonstrating compliance. These obligations effectively impose additional production costs. The administrative burden is also introduced to US firms exporting services to the EU (who operate under a self-certified framework for compliance), as they are assumed to follow EU rules. In Bauer et al., this burden is estimated to be less than 0.6 % of the export value, which is a low estimation, given the costs of pre-existing regulatory divergences and incompabilities are expected to be up to 14.9 % of the export value in services.[20]

In an upper-bound scenario, the study assumes that all non-compliant, foreign-based providers (i. e. outside the jurisdiction of the GDPR) will be shut off from transferring data. This effectively leads to a data localisation requirement, and as a result, non-EU service exporters must acquire data-processing capacities inside the EU. Countries that the EU deemed as having 'adequate' data protection (which were accountable for only 6 % of global services trade) may still transfer data freely in the same manner as those within the Single Market,[21] while most major economies are effectively cut off. This is expected to result in a price increase of between 4 and 13 % on services originating from the US—depending on sector—or between 7 and 26 % on services from the rest of the world.

## 5.4 Estimation of the Economic Impact

Taken together, the EU and the US account for half of world's GDP, so the impact of GDPR is most clearly visible on EU and US trade. In the lower-bound scenario, the trade barriers on primarily US-based suppliers lead to a drop of at least −0.2 % in their services exports to the EU. This negative impact most likely represents small-sized enterprises that are displaced from the market by the increased trade barriers and have little means to establish subsidiaries or purchase data processing

[18] Narayanan et al. (2012).

[19] See note 15, United Kingdom (2012).

[20] Francois (2013).

[21] Andorra, Argentina, Canada, Faroe Islands, Guernsey, Iceland, Isle of Man, Israel, Jersey, Lichtenstein, New Zealand, Norway, Switzerland and Uruguay.

capabilities inside the EU.[22] EU services exports to the US are also severely affected in the first scenario. With a lesser degree of competition leading to a higher level of input costs, and, in turn, further lost competitiveness, EU services exports to the US would also decrease by at least 0.6 %. As if often the case in trade economics, these numbers may seem minuscule and marginal. But in comparison, an ambitious free trade agreement (FTA) designed to liberate trade between its counterparts, such as the EU–US Transatlantic Trade and Investment Partnership (TTIP), is estimated to increase bilateral trade by approximately 0.7 %,[23] and the loss from GDPR offsets a significant share of the benefits.

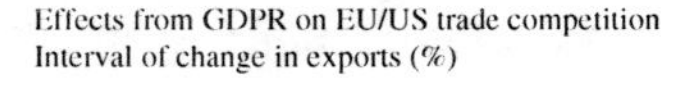


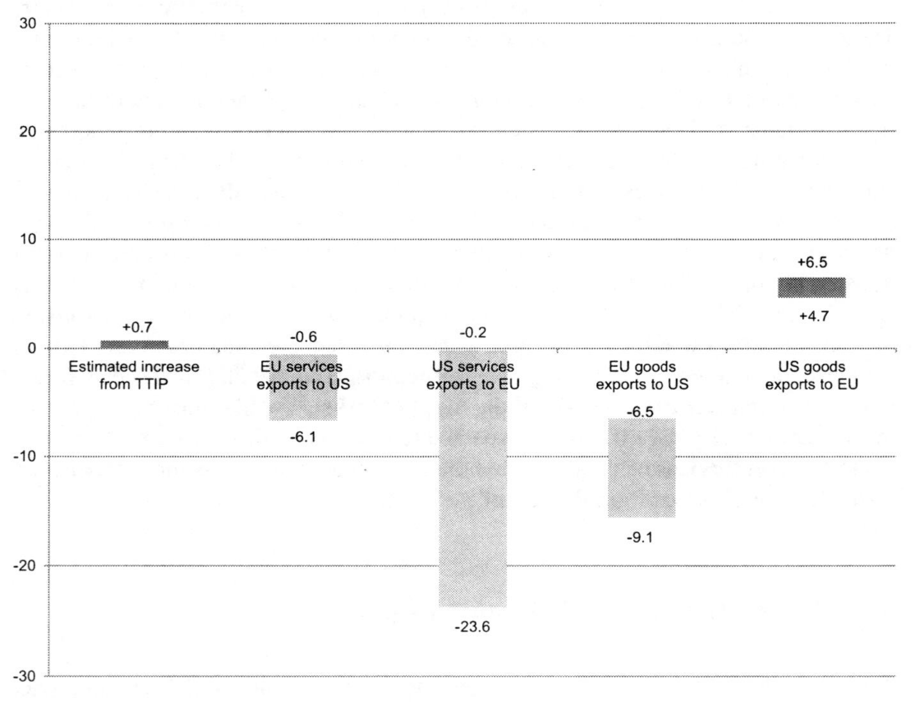


Francois (2013); Bauer, Erixon, Krol, Lee-Makiyama & Verschelde (2013)

In the upper-bound scenario, it is assumed that a new regulation will forcefully relocate data processing in third countries to either the EU or one of the jurisdictions deemed to have an 'adequate' legislation.[24] In this scenario, the US decreases

[22] Multinational corporations commonly employ legal constructs like model contract clauses (MCCs) or binding corporate rules (BCRs).

[23] See note 20.

[24] See note 17.

its services exports to the EU by up to 24%. Exporters from other countries that are deemed as 'non-adequate' see an even larger decrease of up to 80% in services exported to the EU. The 'adequate' economies will in part fill the void left by their competitors and increase their exports into the EU by up to 21%. Measured in total sector output, the European services sectors will decrease all activities by up to 2.7%, while EU-equivalent countries will increase their activities by 0.6%.

Lower imports into the Single Market do not necessarily translate to increased global competitiveness for those who are based in the EU. Any restrictions on cross-border data flows and services supply-chains would seriously affect Europe's ability to export outside its borders. EU services exports to the US are significantly hampered in comparison with previous scenarios (−6.7%). As mentioned above, the model only affects manufacturing sectors through indirect effects from higher input costs of services (logistics, financing, retailing, business services etc.). However, these indirect effects are sufficient to reduce EU goods exports to the US by up to 9.1%.

In sum, the effects of reduced trade (in both services and manufacturing) and decreased efficiencies leave a substantial imprint on EU GDP—a loss of between 0.3% (the lower-bound compliance cost scenario) and 1.3% (the upper-bound scenario) if full restrictions on cross-border data flows are implemented unilaterally by the EU. This figure is equivalent to 1–3 times the rate of economic slowdown during the euro crisis. In terms of direct welfare effects on the EU—i. e., the accumulative effect on utility change—are equivalent to a loss of 52–170 billion $ (€ 40–131 billion), or up to 338 USD (€260) per European citizen, or 1,353 $ (€ 1,041) for a household of four people.

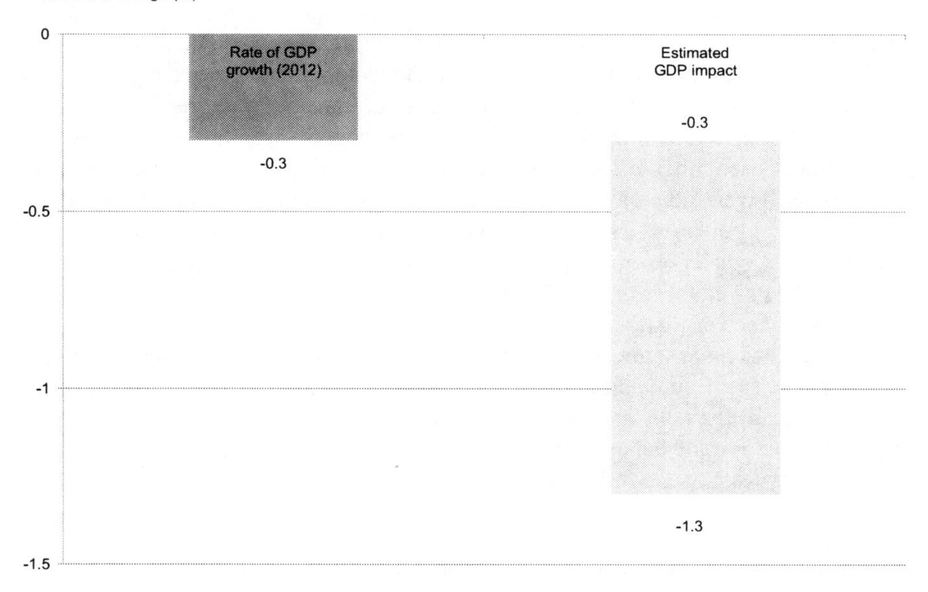


Eurostat (2013); Bauer, Erixon, Krol, Lee-Makiyama & Verschelde (2013)

## 5.5 Conclusion—the Redistributive Change of GDPR

The model described does not take into account the economic gains from harmonisation and increased consumer confidence—benefits that inarguably exist, despite the problems of estimating them. In order to offset the negative effects from productivity and trade losses identified in the model, harmonization must lead to a consistent increase of all private final consumption in the EU by 13 % on all goods and services, whereas historically, consumption has varied between −2 % and +2.6 % per year since 2004.[25]

As became evident in the previous section, the composite effect from the shifts in international trade would be considerable for the EU and its trading partners alike. By increasing the input cost for the services industry, the regulation works in effect as a redistribution from the general services industry (that accounts for 71.5 % of employment),[26] to a very few individuals. Given that the main beneficiaries inside the Single Market are those who enjoy income increases from higher prices (data processors, mainly in low-cost member states), those who gain economically from GDPR are counted in fractions of a percent. Furthermore, the measure disproportionately affects EU manufacturing exports through the loss of relative competitiveness with the rest of the world, in particular advanced competitors like the US.

For other economies, the net gains are concentrated to the US manufacturing sector that reaps the benefits of weaker competition, and the non-EU countries that are deemed as 'adequate', who see a very marginal production output in services. These gains do not compensate for welfare losses in the EU. The aggregated redistribution results in a net loss in the world economy, with demand in Europe that will never be replaced by new sources of supply.

As a conclusion, restrictions on free movement of data in a service-oriented and trade-dependent economy affects the protecting market more than those who are restricted in accessing it. This is largely due to the fact that data processing generates a minuscule amount of employment relative to its usage. The degree of economic loss incurred is also a function of the price increase—i. e., the extent to which competition (or number of foreign suppliers) becomes limited. In order to minimise negative spillover and maximise efficiency, the absolute number and variety of input suppliers at different price points should be maintained. In an open economy, this generally entails opening up to foreign supply at world market prices—and in the case of data-related services, the price differential between foreign and domestic or EU supply is substantial.

The findings have some bearing on the discussions around policy design and efficiency. The first point, and perhaps the most obvious one, concerns the effectiveness of mercantilist policies, i. e. the notion of increasing national wealth through a maximisation of trade balances (usually by imposing trade barriers), providing artificial economies of scale to the domestic suppliers, and thereby strengthening their

[25] Using final private consumption with constant prices (2005), Eurostat (2014).

[26] Eurostat (2013).

export competitiveness. This analysis suggests that data restrictions as import barriers are counterproductive, a conclusion consistent with international trade theory and the concept of global value chains.[27] Mercantilism has ceased to work for even a well-diversified economy like the EU, which enjoys the world's largest internal market and a large number of suppliers.

Secondly, the political economy of gains and losses proscribes that the economic beneficiaries should not be able to uphold the data restrictions as a commercial policy that unilaterally puts Europe at a relative loss to other economies. As we have also seen, the possibilities for offsetting the negative impact through trade policy instruments (such as FTAs) are relatively limited. Regulations themselves that restrict data flows may be a subject of negotiation and subsequent deregulation as a consequence of such agreements. The EU itself has negotiated free movement of data in its FTAs since its 2011 agreement with the Republic of Korea.[28] One caveat in this equation is the economic loss suffered by others independently of the EU legislation—for example, US business estimates that it stands to lose up to $ 35 billion on cloud business from loss of consumer confidence.[29]

The third point concerns the choices available to policymaking, proviso the restraints set by the first two—namely that data connectivity and economic interdependency have limited the policy space for even seemingly non-economic regulations, in the same manner that globalisation has shrunk the policy space on corporate taxation, tariffs or monetary policy.[30] Negative cross-border externalities (whether they are data privacy breaches or toxic emissions) tend to lead to multilateral or intergovernmental governance as unilateral legislation fails to address externalities created outside the national jurisdiction. However, international standard setting and governance on data privacy has failed to materialise, given the differences in constitutional and regulatory structure.

A final point concerns the incentives behind data restrictions according to positive political theory. If the earlier points on international political economy are valid, data restrictions and its losses for Europe seem to contradict the objectives of the EU as a market liberal and post-modern construct.[31] This contradiction can have several explanations—from the assumptions of public choice theory (which concerns the incentives that shape the political institutions in legislative processes) the negative redistribution of wealth is secondary to the gains from redistribution of internal competences, primarily between the European Commission and the EU member states. An alternative interpretation to public choice is the theory of public interest, which stipulates that the regulation creates a general benefit for the society at large, by benevolent public servants overriding minority interest groups. However, since the societal costs are applied to a vast majority of interest groups (consumers, manufacturing and services producers), these general benefits must be

[27] See Baldwin (2012), Miroudot ((2012).

[28] See EU–South Korea Free Trade Agreement, Art. 7.43.

[29] Castro (2013).

[30] See Mayer (2009), Brown and Stern (2006), Cooper (1968).

[31] Erixon (2009).

presented to an even larger group than the costs, and of considerable size. According to the objectives of the public choice theory, the efficiency of the GDPR as a regulation is either neutral or even dependent on imposing economic harm to self and others. In the case of public interest, the minimisation of negative externalities by imposing the least-trade restrictive measure does not contradict the objectives of the regulation. The determinants of the policy design of GDPR should be determined by this conclusion.

## References


Baldwin, R. 2012. Global manufacturing value chains and trade rules. World Economic Forum.

Bauer, Erixon, Lee-Makiyama, Krol, and Bert Verschelde. 2013. The economic importance of getting data protection right. European Centre for International Political Economy (ECIPE).

Brown, A. G., and R. M. Stern. 2006. Global market integration and national sovereignty. World Economy.

Castro, D. 2013 August. How much will PRISM cost the U.S. Cloud computing industry?, ITIF.

Cooper, R. N. 1968. The economics of interdependence: Economic policy in the Atlantic community. Council on foreign relations.

Erixon, F. 2009. Twilight of soft mercantilism: Europe and foreign economic power. Europe's world, 13/2009.

European Commission. On the protection of individuals with regard to the processing of personal data and on the free movement of such data. Directive 95/46/EC.

European Commission. Impact assessment, regulation of the European parliament and of the council on the protection of individuals with regard to the processing of personal data and on the free movement of such data (General Data Protection Regulation) and Directive of the European parliament and of the council on the protection of individuals with regard to the processing of personal data by competent authorities for the purposes of prevention, investigation, detection or prosecution of criminal offences or the execution of criminal penalties, and the free movement of such data, SEC(2012) 72 final.

European Union. 2000. Charter of fundamental rights of the European Union, 2000/C 364/01.

European Union. Treaty on the functioning of the European Union (TFEU).

Floridi, L. 2013. The onlife manifesto: Being human in a hyperconnected era, (Introduction).

Francois, J. 2013. Reducing transatlantic barriers to trade and investment: An economic assessment. CEPR.

Lee-Makiyama, H. 2013. Testimony to the US international trade commission. Digital trade in the U.S. and global economies. Investigation no. 332–531.

Mayer, J. 2009. Policy space: What, for what, and where? Development Policy Review.

McKinsey Global Institute. Internet matters: The net's sweeping impact on growth, jobs, and prosperity.

Miroudot, Sebastian. 2012. Trade policy implication of global value chains: Scoping paper, OECD.

Narayanan, G., Angel A. Badri, and R. McDougall, eds. 2012. Global trade, assistance, and production: The GTAP 8 data base. Center for Global Trade Analysis.

OECD. 2005. Input to the United Nations working group on internet governance, DSTI/ICCP(2005)4.

World Bank. 2013. World Bank Data.

World Trade Organization. General agreement on trade in services (GATS).

UNCTAD. 2009. Information Economy Report.

United Kingdom. 2012. Ministry of justice, impact assessment on proposal for EU data protection regulation.

# Chapter 6
# The Rise of the MASs

**Luciano Floridi**

**Abstract** The post-Westphalian Nation State developed by becoming more and more an Information Society. However, in so doing, it progressively made itself less and less *the* main information agent, because what made the Nation State possible and then predominant, as a historical driving force in human politics, namely ICTs, is also what is now making it less central, in the social, political and economic life of humanity across the world. ICTs fluidify the topology of politics. They do not merely enable but actually promote (through management and empowerment) the agile, temporary and timely aggregation, disaggregation and re-aggregation of distributed groups around shared interests across old, rigid boundaries represented by social classes, political parties, ethnicity, language barriers, physical barriers, and so forth. This is generating a new tension between the Nation State, still understood as a major organisational institution, yet no longer monolithic but increasingly morphing into a multiagent system itself, and a variety of equally powerful, indeed sometimes even more politically influential and powerful, non-Statal organisations. Geo-politics is now global and increasingly non-territorial, but the Nation State still defines its identity and political legitimacy in terms of a sovereign territorial unit, as a Country. Such tension calls for a serious exercise in conceptual re-engineering: how should the new informational multiagent systems (MASs) be designed in such a way as to take full advantage of the socio-political progress made so far, while being able to deal successfully with the new global challenges (from the environment to the financial markets) that are undermining the legacy of that very progress? In the lecture, I shall defend an answer to this question in terms of a design of political MAS based on principles borrowed from information ethics.

L. Floridi (✉)
Oxford Internet Institute, University of Oxford,
St Giles 1, OX1 3JS, Oxford, UK
e-mail: luciano.floridi@oii.ox.ac.uk

L. Floridi (ed.), *Protection of Information and the Right to Privacy – A New Equilibrium?*,
Law, Governance and Technology Series 17, DOI 10.1007/978-3-319-05720-0_6,

## 6.1 From History to Hyperhistory

More people are alive today than ever before in the evolution of humanity. And more of us live longer and better today than ever before. To a large measure, we owe this to our technologies, at least insofar as we develop and use them intelligently, peacefully, and sustainably.

Sometimes, we may forget how much we owe to flakes and wheels, to sparks and ploughs, to engines and satellites. We are reminded of such deep technological debt when we divide human life into prehistory and history. That significant threshold is there to acknowledge that it was the invention and development of information and communication technologies (ICTs) that made all the difference between who we were and who we are. It is only when the lessons learnt by past generations began to evolve in a Lamarckian rather than a Darwinian way that humanity entered into history.

History has lasted 6,000 years, since it began with the invention of writing in the fourth millennium BC. During this relatively short time, ICTs have provided the recording and transmitting infrastructure that made the escalation of other technologies possible. ICTs became mature in the few centuries between Guttenberg and Turing. Today, we are experiencing a radical transformation in our ICTs that could prove equally significant, for we have started drawing a new threshold between history and a new age, which may be aptly called *hyperhistory*. Let me explain.

Prehistory and history work like adverbs: they tell us *how* people live, not *when* or *where*. From this perspective, human societies currently stretch across three ages, as ways of living. According to reports about an unspecified number of uncontacted tribes in the Amazonian region, there are still some societies that live prehistorically, without ICTs or at least without recorded documents. If one day such tribes disappear, the end of the first chapter of our evolutionary book will have been written. The greatest majority of people today still live historically, in societies that rely on ICTs to *record* and *transmit* data of all kinds. In such historical societies, ICTs have not yet overtaken other technologies, especially energy-related ones, in terms of their vital importance. There are then some people around the world who are already living hyperhistorically, in societies or environments where ICTs and their data *processing* capabilities are the necessary condition for the maintenance and any further development of societal welfare, personal well-being, as well as intellectual flourishing. The nature of conflicts provides a sad test for the reliability of this tripartite interpretation of human evolution. Only a society that lives hyperhistorically can be vitally threatened informationally, by a cyber attack. Only those who live by the digit may die by the digit.

To summarise, human evolution may be visualised as a three-stage rocket: in prehistory, there are no ICTs; in history, there are ICTs, they record and transmit data, but human societies depend mainly on other kinds of technologies concerning primary resources and energy; in hyperhistory, there are ICTs, they record, transmit and, above all, process data, and human societies become vitally dependent on them and on information as a fundamental resource. If all this is even approximately correct, humanity's emergence from its historical age represents one of the most significant steps it has ever taken. It certainly opens up a vast horizon of opportunities as well as challenges and difficulties, all essentially driven by the recording, transmitting, and processing powers of ICTs. From synthetic biochemistry to neuroscience, from the Internet of things to unmanned planetary explorations, from green technologies to new medical treatments, from social media to digital games, from agricultural to financial applications, from economic developments to the energy industry, our activities of discovery, invention, design, control, education, work, socialisation, entertainment, care, security, business and so forth would be not only unfeasible but unthinkable in a purely mechanical, historical context. They have all become hyperhistorical in nature.

### *6.1.1 Political Apoptosis*

In 2011, the total world wealth[1] was calculated to be $ 231 trillion, up from $ 195 trillion in 2010.[2] Since we are almost 7 billion, that was about $ 33,000 per person, or $ 51,000 per adult, as the report indicates. The figures give a clear sense of the level of inequality. In the same year, we spent $ 498 billion on advertisements.[3] Perhaps for the first time, we also spent more on ways to entertain ourselves than on ways to kill each other. The military expenditure in 2010 was $ 1.74 trillion,[4] and that on entertainment and media was expected to be around $ 2 trillion, with digital entertainment and media share growing to 33.9% of all spending by 2015, from 26% in 2011.[5] Meanwhile, we spent $ 6.5 trillion (this is based on 2010 data) on fighting health problems and premature death, much more than the military and the entertainment and media budgets put together. All these trillions were closely linked and often overlapped with the budget for Information and Communication Technologies (ICTs), on which we spent $ 3 trillion in 2010.[6] We can no longer unplug our world from ICTs without turning it off.

Hyperhistory, and the evolution of the infosphere in which we live, are quickly detaching future generations from ours. Of course, this is not to say that there is no continuity, both backwards and forwards. *Backwards*, because it is often the case that the deeper a transformation is, the longer and more widely rooted its causes may be. It is only because many different forces have been building the pressure for a long time that radical changes may happen all of a sudden, perhaps unexpectedly. It is not the last snowflake that breaks the branch of the tree. In our case, it is

[1] This is defined as the value of financial assets plus real assets (mainly housing) owned by individuals, less their debts.

[2] Source: *The Credit Suisse Global Wealth Report 2011*, available online.

[3] Source: *Nielsen Global AdView Pulse Q4 2011*, available online.

[4] Source: Stockholm International Peace Research Institute, *Military Expenditure Database*, available online.

[5] Source: PricewaterhouseCoopers, *Global Entertainment and Media Outlook 2007–2011*, available online.

[6] Source: IDC, *Worldwide IT Spending Patterns: The Worldwide Black Book*, available online.

certainly history that begets hyperhistory. There is no ASCII without the alphabet. *Forwards*, because we should expect historical societies to survive for a long time in the future. Despite globalisation, human societies do not parade uniformly forward, in neat and synchronised steps.

We are witnessing a slow and gradual process of political *apoptosis*. Apoptosis, also known as programmed cell death, is a natural and normal form of self-destruction in which a programmed sequence of events leads to the self-elimination of cells. Apoptosis plays a crucial role in developing and maintaining the health of the body. One may see this as a natural process of renovation. Here, I am using the expression 'political apoptosis' in order to describe the gradual and natural process of renovation of sovereign states[7] as they develop into information societies. Let me explain.

Simplifying and generalising, a quick sketch of the last 400 years of political history in the Western world may look like this. The Peace of Westphalia (1648) meant the end of World War Zero, namely the Thirty Years' War, the Eighty Years' War, and a long period of other conflicts during which European powers, and the parts of the world they controlled, massacred each other for economic, political and religious reasons. Christians brought hell to each other, with staggering violence and unspeakable horrors. The new system that emerged in those years, the so-called *Westphalian order*, saw the coming of maturity of sovereign states and then national states as we still know them today; France, for example. Think of the time between the last chapter of *The Three Musketeers*—when D'Artagnan, Aramis, Porthos, and Athos take part in Cardinal Richelieu's siege of La Rochelle in 1628—and the first chapter of *Twenty Years Later*, when they come together again, under the regency of Queen Anne of Austria (1601–1666) and the ruling of Cardinal Mazarin (1602–1661).

The state did not become a monolithic, single-minded, well-coordinated entity. It was not the sort of beast that Hobbes described in his *Leviathan*, nor the sort of robot that a later, mechanical age would incline us to imagine. But it did rise to the role of the binding power, the system able to keep together and influence all the different agents comprising it, and coordinate their behaviours, as long as they were falling within the scope of its geographical borders. These acquired the metaphorical status of a state's skin. States became the independent agents that played the institutional role in a system of international relations. And the principles of sovereignty (each state has the fundamental right of political self-determination), legal equality (all states are equal), and non-intervention (no state should interfere with the internal affairs of another state) became the foundations of such a system of international relations.

Citizenship had been discussed in terms of biology (your parents, your gender, your age…) since the early city-states of ancient Greece. It became more flexi-

[7] Using standard vocabulary, by nation I refer to a socio-cultural entity comprising people united by language and culture. By state, I refer to a political entity that has a permanent population, a defined territory, a government, and the capacity to enter into relations with other states (Montevideo Convention 1933). The kurds are a typical example of a nation without a state.

ble (types of citizenship) when it was conceptualised in terms of legal status as well. This was the case under the Roman Empire, when acquiring a citizenship—a meaningless idea in purely biological contexts—meant becoming a holder of rights. With the modern state, geography started playing an equally important role, mixing citizenship with language, nationality, ethnicity, and locality. In this sense, the history of the passport is enlightening. As a means to prove the holder's identity, it is acknowledged to be an invention of King Henry V of England (1386–1422), a long time before the Westphalian order took place. However, it was the Westphalian order that transformed the passport into a document that entitles the holder not to travel (because a visa may also be required, for example) or be protected abroad, but to return to the country that issued the passport. The passport became like an elastic band that ties the holder to a geographical point, no matter how far in space and prolonged in time the journey in other lands has been. Such a document became increasingly useful the better that geographical point was defined. Travelling was still quite passport-free in Europe until the First World War. Only then did security pressure and techno-bureaucratic means catch up with the need to disentangle and manage all those elastic bands travelling around by means of a new network, the railway.

Back to the Westphalian order. Now that the physical and legal spaces overlap, they can both be governed by sovereign powers, which exercise control, impose laws, and ensure their respect by means of physical force within the state's borders. Geographical mapping is not just a matter of travelling and doing business, but also an inward-looking question of controlling one's own territory, and an outward-looking question of positioning oneself on the globe. The taxman and the general look at those geographical lines with eyes very different from those of today's users of Expedia. For sovereign states act as agents that can, for example, raise taxes within their borders and contract debts as legal entities (hence our current terminology in terms of 'sovereign bonds', for example, which are bonds issued by national governments in foreign currencies), and of course dispute borders, often violently. Part of the political struggle becomes not just a silent tension between different components of the state as a multiagent system, say the clergy vs. the aristocracy, but an explicitly codified balance between the different agents constituting it. In particular, Montesquieu (1689–1755) suggests the classic division of the state's political powers that we take for granted today: a legislature, an executive, and a judiciary. The state as a multiagent system organises itself as a network of these three 'small worlds', among which only some specific channels of information are allowed. Today, we may call that arrangement Westphalian 2.0.

With the Westphalian order, modern history becomes the age of the state. The state arises as *the* information agent, which legislates on, and at least tries to control, the technological means involved in the information life-cycle, including education, census,[8] taxes, police records, written laws, press, and intelligence. Already most

[8] The Latin word means "estimate". Already the Romans, who were well aware of the importance of information and communication in such a large empire for administrative and taxing purposes, carried out a census every 5 years.

of the adventures in which D'Artagnan is involved are caused by some secret communication.

As the information agent, the state fosters the development of ICTs as a means to exercise and maintain legal force, political power, and social control, especially at times of international conflicts, frequent unrests, and fragile peace. For example, in 1790–1795, during the French Revolution, the French government needed a system of speedy communication to receive intelligence and transmit orders in time to counterbalance the hostile manoeuvres of the allied forces that surrounded France: Britain, the Netherlands, Prussia, Austria, and Spain. To satisfy such need, Claude Chappe (1763–1805) invented the first system of telegraphy (he actually coined the word 'telegraph'). It consisted of mechanical semaphores that could transmit messages across the country in a matter of hours. It became so strategic that when Napolen begun preparations to resume war in Italy, in 1805, he ordered a new extension from Lyon to Milan. At its peak, the *Chappe* telegraph was a network of 534 stations, covering more than 5,000 km (3,106 miles). The reader may remember its crucial appearance in Alexandre Dumas' *The Count of Monte Cristo* (1844) where the Count bribes an operator to send a false message to manipulate the financial market to his own advantage. In fictional as in real life, whoever controls information controls the issuing events.

Through the centuries, the state moves from being conceived as the ultimate guarantor and defender of a *laisser-faire* society to a Bismarckian welfare system, which takes full care of its citizens. In both cases, the state remains the primary collector, producer, and controller of information. However, by fostering the development of ICTs, the state ends undermining its own future as the only, or even the main, information agent. This is the political apoptosis I mentioned above. For in the long run, ICTs contribute to transforming the state in an information society, which makes possible other, sometimes even more powerful information agents, which may determine political decisions and events. And so ICTs help shift the balance against centralised government, in favour of distributed governance and international, global coordination.

The two World Wars are also clashes of sovereign states resisting mutual coordination and inclusion as part of larger multiagent systems. The *Bretton Woods* conference may be interpreted as the event that seals the beginning of the political apoptosis of the state. The gathering in 1944 of 730 delegates from all 44 Allied nations at the Mount Washington Hotel in Bretton Woods, New Hampshire, United States, regulated the international monetary and financial order after the conclusion of Second World War. It saw the birth of the International Bank for Reconstruction and Development (this, together the International Development Association, is now known as the World Bank), of the General Agreement on Tariffs and Trade (GATT, replaced by the World Trade Organisation in 1995), and the International Monetary Fund. In short, Bretton Woods brought about a variety of multiagent systems as supranational or intergovernmental forces involved with the world's political, social, and economic problems. These and similar agents became increasingly powerful and influential, as the emergence of the *Washington Consensus* clearly indicated.

John Williamson[9] coined the expression 'Washington Consensus' in 1989. He used it in order to refer to a set of ten, specific policy recommendations, which, he argued, constituted a standard strategy adopted and promoted by institutions based in Washington, D.C.—such as the US Treasury Department, the International Monetary Fund, and the World Bank—when dealing with countries coping with economic crises. The policies concerned macroeconomic stabilization, economic opening with respect to both trade and investment, and the expansion of market forces within the domestic economy. In the past quarter of a century, the topic has been the subject of intense and lively debate, in terms of correct description and acceptable prescription. Like the theory of a Westphalian doctrine I outlined above, the theory of a Washington Consensus is not devoid of problems. Does the Washington Consensus capture a real historical phenomenon? Does the Washington consensus ever achieve its goals? Is it to be re-interpreted, despite Williamson's quite clear definition, as the imposition of neoliberal policies by Washington-based international financial institutions on troubled countries? These are important questions, but the real point of interest here is not the interpretative, economic, or normative evaluation of the Washington Consensus. Rather, it is the fact that the very idea, even if it remains only an influential idea, captures a significant aspect of our hyperhistorical, post-Westphalian time. The Washington Consensus is a coherent development of Bretton Woods. Both highlight the fact that, after the Second World War, organisations and institutions (not only those in Washington D.C.) that are not states but rather non-governmental multiagent systems, are openly acknowledged to act as major, influential forces on the political and economic scene internationally, dealing with global problems through global policies. The very fact that the Washington Consensus has been accused (no matter whether correctly or not) of disregarding local specificities and global differences reinforces the point that a variety of powerful multiagent systems are now the new sources of policies in the globalised information societies. As a final reminder, let me mention a rather controversial report, entitled *Top 200: The Rise of Corporate Global Power*. It offered some years ago an analysis of corporate agents.[10] Perhaps the most criticised part was a comparison between countries' yearly GDP and companies' yearly sales (revenues or turnover). Despite this potential shortcoming, it still makes for interesting reading. According to the report:

of the 100 largest economies in the world, 51 are (were, in 2000) corporations; only 49 are (were, in 2000) countries.

The criticism remains, but the percentage has probably moved in favour of the number of companies, and what represents a unifying unit of comparison is that both GDP and revenues buy you clout. When multiagent systems of such dimensions take decisions, their effects are deep and global.

Today, we know that global problems—from the environment to the financial crisis, from social justice to intolerant religious fundamentalisms, from peace to health conditions—cannot rely on sovereign states as the only source of a solution

[9] Williamson (1993, pp. 1329–1336).

[10] Anderson and Cavanagh (2000).

because they involve and require global agents. However, there is much uncertainty about the design of the new multiagent systems that may shape humanity's future. Hyperhistorical societies are post-Westphalian, because of the emergence of the sovereign state as the modern, political information agent. They are post-Bretton Woods, because of the emergence of non-state multiagent systems as hyperhistorical players in the global economy and politics. This helps explain why one of the main challenges faced by hyperhistorical societies is how to design the right sort of multiagent systems. These systems should take full advantage of the socio-political progress made in modern history, while dealing successfully with the new global problems, which undermine the legacy of that very progress, in hyperhistory.

### 6.1.2 A New Informational Order?

The shift from a historical, Westphalian order to a post-Bretton Woods, hyperhistorical predicament in search of a new equilibrium may be explained by many factors. Four are worth highlighting in the context of this book.

First, *power*. ICTs 'democratise' data and the processing/controlling power over them, in the sense that now both tend to reside and multiply in a multitude of repositories and sources. Thus, ICTs can create, enable, and empower a potentially boundless number of non-state agents, from the single individual to associations and groups, from macro-agents, like multinationals, to international, intergovernmental as well as nongovernmental, organisations and supranational institutions. The state is no longer the only, and sometimes not even the main, agent in the political arena that can exercise informational power over other informational agents, in particular over human individuals and groups. The European Commission, for example, recognised the importance of such new agents in the *Cotonou Agreement* between the European Union (EU) and the Africa, Caribbean and Pacific (ACP) countries, by acknowledging the important role exercised by a wide range of non-governmental development actors, and formally recognising their participation in ACP-EU development cooperation. According to Article 6 of the *Cotonou Agreement*, such non-state actors comprise:

> the private sector; economic and social partners, including trade union organisations; civil society in all its forms, according to national characteristics.

The 'democratisation' brought about by ICTs is generating a new tension between power and force, where power is informational, and exercised through the elaboration and dissemination of norms, whereas force is physical, and exercised when power fails to orient the behaviour of the relevant agents and norms need to be enforced. Note that the more physical goods and even money become information-dependent, the more the informational power exercised by multiagent systems acquires a significant financial aspect.

Second, *geography*. ICTs de-territorialise human experience. They have made regional borders porous or, in some cases, entirely irrelevant. They have also created, and are exponentially expanding, regions of the infosphere where an increasing number of agents, not necessarily only human, operate and spend more and more time, the onlife experience. Such regions are intrinsically stateless. This is generating a new tension between geo-politics, which is global and non-territorial, and the state, which still defines its identity and political legitimacy in terms of a sovereign territorial unit, as a country.

Third, *organisation*. ICTs fluidify the topology of politics. They do not merely enable but actually promote, through management and empowerment, the agile, temporary and timely aggregation, disaggregation and re-aggregation of distributed groups 'on demand', around shared interests, across old, rigid boundaries, represented by social classes, political parties, ethnicity, language barriers, physical barriers, and so forth. This is generating new tensions between the state, still understood as a major organisational institution, yet no longer rigid but increasingly morphing into a flexible multiagent system itself, and a variety of equally powerful, indeed sometimes even more powerful and politically influential (with respect to the old sovereign state) non-state organisations, the other multiagent systems on the block. Terrorism, for example, is no longer just a problem concerning internal affairs—consider forms of terrorism in the Basque Country, Germany, Italy, or Northern Ireland—but also an international confrontation with a distributed, multiagent system such as Al-Qaeda.

Finally, *democracy*. Changes in power, geography, and organisation, reshape the debate on democracy, the oldest and safest form of power crowdsourcing. We used to think that, ideally, democracy should be a direct and constant involvement of all citizens in the running of their society and its business, their *res publica*. Direct democracy, if feasible, was about how the state could re-organise itself internally, by designing rules and managing the means to promote forms of negotiation, in which citizens could propose and vote on policy initiatives directly and almost in real time. We thought of forms of direct democracy as complementary options for forms of representative democracy. It was going to be a world of 'politics always-on'. The reality is that direct democracy has turned into a mass media-ted democracy, in the ICT sense of new social media. In such digital democracies, distributed groups, temporary and timely aggregated around shared interests, have multiplied and become sources of influence external to the state. Citizens vote for their representatives but can constantly influence them via opinion polls almost in real time. Consensus-building has become a constant concern based on synchronic information.

Because of the factors just analysed—power, geography, organisation, and democracy—the unique position of the historical state as *the* information agent is being undermined from below and overridden from above. Other multiagent systems have the data, the power and sometimes even the force—as in the different cases of the UN, of groups' cyber threats, or of terrorist attacks—the space, and the organisational flexibility to erode the modern state's political clout. They can appropriate some of its authority and, in the long run, make it redundant in contexts where it was once the only or the predominant informational agent. The Greek economic crisis, which began in late 2009, offers a good example. The Greek government and the Greek state had to interact 'above' with the EU, the European Central Bank, the International Monetary Fund, the rating agencies, and so forth. They had to interact 'below' with the Greek mass media and the people in Syntagma Square, the financial markets and international investors, German public opinion, and so forth. Because the state is less central than in the nineteenth century, countries such as Belgium and Italy may work fine even during long periods without governments or when governed by dysfunctional ones, on 'automatic pilot'.

A much more networked idea of political interactions makes possible a degree of tolerance towards, and indeed feasibility of localisms, separatisms, as well as movements and parties favouring autonomy or independence that would have been unacceptable by modern nation states, which tended to encourage aggregating forms of nationalism but not regionalism. From Padania (Italy) to Catalonia (Spain), from Scotland (Great Britain) to Bavaria (Germany), one is reminded that, in almost any European country, hyperhistorical trends may resemble pre-Westphalian equilibria among a myriad of regions. The long 'list of active separatist movements in Europe' in Wikipedia is both informative and eye opening. Unsurprisingly, the Assembly of European Regions (originally founded as the Council of the Regions of Europe in 1985), which brings together over 250 regions from 35 countries along with 16 interregional organisations, has long been a supporter of subsidiarity, the decentralising principle according to which political matters ought to be dealt with by the smallest, lowest, or least centralised authority that could address them effectively.

Of course, the historical state is not giving up its role without a fight. In many contexts, it is trying to reclaim its primacy as the information super-agent governing the political life of the society that it organises.

In some cases, the attempt is blatant. In the UK, the Labour Government introduced the first Identity Cards Bill in November 2004. After several intermediary stages, the Identity Cards Act was finally repealed by the Identity Documents Act 2010, on 21 January 2011. The failed plan to introduce compulsory ID in the UK should be read from a modern perspective of preserving a Westphalian order.

In many other cases, it is 'historical resistance' by stealth, as when an information society is largely run by the state. In this case, the state maintains its role of major informational agent no longer just legally, on the basis of its power over legislation and its implementation, but also economically, on the basis of its power over the majority of information-based jobs. The intrusive presence of so-called State Capitalism with its State Owned Enterprises all over the world, from Brazil, to France, to China, is a symptom of hyperhistorical anachronism.

Similar forms of resistance seem only able to delay the inevitable rise of political multiagent systems. Unfortunately, they may involve not only costs, but also huge risks, both locally and globally. Recall that the two World Wars may be seen as the end of the Westphalian system. Paradoxically, while humanity is moving into a hyperhistorical age, the world is witnessing the rise of China, currently a most 'historical' state, and the decline of the US, a state that more than any other superpower in the past already had a hyperhistorical and multiagent vocation in its federal organisation. We might be moving from a Washington Consensus to a *Beijing Consensus* described by Williamson as consisting of incremental reform, innovation and experimentation, export-led growth, state capitalism, and authoritarianism.[11] This is risky, because the anachronistic historicism of some of China's policies and humanity's growing hyperhistoricism are heading towards a confrontation. It may not be a conflict, but hyperhistory is a force whose time has come, and while it seems likely that it will be the Chinese state that will emerge deeply transformed, one can only hope that the inevitable friction will be as painless and peaceful as possible. The financial and social crises that the most advanced information societies are undergoing may actually be the painful but still peaceful price we need to pay to adapt to a future post-Westphalian system.

The previous conclusion holds true for the historical state in general. In the future, political multiagent systems will acquire increasing prominence, with the problem that the visibility and transparency of such acquisition of power may be rather unclear. It is already difficult to monitor and understand politics when states are the main players. It becomes even harder when the agents in question have fuzzier features, more opaque behaviours, and are much less easily identifiable, let alone accountable. At the same time, it is to be hoped that the state itself will progressively abandon its resistance to hyperhistorical changes and evolve even more into a multiagent system. Good examples are provided by devolution, the transfer of state's sovereign rights to supranational European institutions, or the growing trend in making central banks, like the Bank of England or the European Central Bank, independent, public organisations.

The time has come to consider the nature of political multiagent system more closely and some of the questions that its emergence is already posing.

### *6.1.3 The Political Multiagent System*

A political multiagent system is a single agent, constituted by other systems, which is

a. *teleological*: the multiagent system has a purpose, or goal, which it pursues through its actions;
b. *interactive*: the multiagent system and its environment can act upon each other;
c. *autonomous*: the multiagent system can change its configurations without direct response to interaction, by performing internal transformations to change its states. This imbues the multiagent system with some degree of complexity and independence from its environment; and finally
d. *adaptable*: the multiagent system's interactions can change the rules by which the multiagent system itself changes its states. Adaptability ensures that the multiagent system learns its own mode of operation in a way that depends critically on its experience.

[11] Williamson (2012, pp. 1–16). The expression 'Beijing Consensus' was introduced by Ramo and Foreign Policy Centre (London, England) (2004) but I am using it here in the sense discussed by Williamson and Halper (2010).

The political multiagent system becomes *intelligent* (in the AI sense of "smart") when it implements the previous features efficiently and effectively, minimising resources, wastefulness and errors, while maximising the returns of its actions.

The emergence of intelligent, political multiagent systems poses many serious questions. Some of them are worth reviewing here, even if only quickly: identity, cohesion, consent, social vs. political space, legitimacy, and transparency.

*Identity* Throughout modernity, the state has dealt with the problem of establishing and maintaining its own *identity* by working on the equation between state = nation. This has often been achieved through the legal means of citizenship and the narrative rhetoric of space (the Mother/Father Land) and time (story in the sense of traditions, recurrent celebrations of past nation-building events, etc.). Consider, for example, the invention of mandatory military service during the French Revolution, its increasing popularity in modern history, but then the decreasing number of sovereign states that still impose it nowadays (your author belongs to the last generation that had to serve in the Italian army for 12 months). Conscription transformed waging war from an eminently economic problem—Florentine bankers financed the English kings during the Hundred Years War (1337–1453), for example—into also a legal problem: the right of the state to send its citizens to die on its behalf. It thus made human life the *penultimate* value, available for the ultimate sacrifice, in the name of patriotism: 'for King and Country'. It is a sign of modern anachronism that, in moments of crisis, sovereign states still give in to the temptation of fuelling nationalism about meaningless, *geographical* spots, often some small islands unworthy of any human loss, including the Falkland Islands (UK) or Islas Malvinas (Argentina), the Senkaku (Japan) or Diaoyu (China) islands, and the Liancourt Rocks, also known as Dokdo (South Korea) or Takeshima (Japan).

*Cohesion* The equation between state, nation, citizenship and land/story had the further advantage of providing an answer to a second problem, that of *cohesion*. For the equation answered not only the question of who or what the state is, but also the question of who or what belongs to the state and hence may be subject to its norms, policies, and actions. New political multiagent systems cannot rely on the same solution. Indeed, they face the further problem of having to deal with the decoupling of their political identity and cohesion. The political identity of a multiagent system may be strong and yet unrelated to its temporary and rather loose cohesion, as is the case with the Tea Party movement in the US. Both identity and cohesion of a political multiagent system may be rather weak, as in the international Occupy movement. Or one may recognise a strong cohesion and yet an unclear or weak political identity, as with the population of tweeting individuals and their role during the Arab Spring. Both identity and cohesion of a political multiagent system are established and maintained through information sharing. The land is virtualised into the region of the infosphere in which the multiagent system operates. So memory (retrievable recordings) and coherence (reliable updates) of the information flow enable a political multiagent system to claim some identity and some cohesion, and therefore offer a sense of belonging. But it is, above all, the fact that the boundaries between the online and offline are disappearing, the appearance of the onlife experience, and hence the fact that the virtual infosphere can affect politically the physical space, that reinforces the sense of the political multiagent system as a real agent. If *Anonymous* had only a virtual existence, its identity and cohesion would be much less strong. Deeds provide a vital counterpart to the virtual information flow to guarantee cohesion. Interactions become more fundamental than things, in a way that is coherent with interactability as a criterion of existence and the development of informational identities. With word play, we might say that *ings* (as in interact-*ing*, process-*ing*, network-*ing*, do-*ing*, be-*ing*, etc.) replace *things*.

*Consent* The breaking up of the equation 'political multiagent system = sovereign state, citizenship, land, story, nation' and the decoupling of identity and cohesion in a political multiagent system have a significant consequence. The age-old theoretical problem of how consent to be governed by a political authority arises is being turned on its head. In the historical framework of social contract theory, the presumed default position is that of a legal opt-out. There is some kind (to be specified) of original consent, allegedly given (for a variety of reasons) by any individual subject to the political state, to be governed by the latter and its laws. The problem is to understand how such consent is given and what happens when an agent, especially a citizen, opts out of it (the out-law). In the hyperhistorical framework, the expected default position is that of a social opt-in, which is exercised whenever the agent subjects itself to the political multiagent system conditionally, for a specific purpose. Simplifying, we are moving from being part of the political consensus to taking part in it, and such part-taking is increasingly 'just in time', 'on demand', 'goal-oriented', and anything but stable, permanent or long-term. If doing politics looks increasingly like doing business this is because, in both cases, the interlocutor, the citizen-customer, needs to be convinced to behave in a preferred way every time anew. Loyal membership is not the default position, and needs to be built and renewed around political and commercial products alike. Gathering consent around specific political issues becomes a continuous process of (re)engagement. It is not a matter of limited attention span. The generic complaint that 'new generations' cannot pay sustained attention to political problems any more is ill-founded. They are, after all, the generations that binge-watch TV. It is a matter of motivating interest again and again, without running into an inflation of information (one more crisis, one more emergency, one more revolution, one more…) and political fatigue (how many times do we need to intervene urgently?). Therefore, the problem is to understand what may motivate repeatedly or indeed force agents (again, not just individual human beings, but all kinds of agents) to give such consent and become engaged, and what happens when such agents, unengaged by default (note, not disengaged, for disengagement presupposes a previous state of engagement), prefer to stay away from the activities of the political multiagent system, inhabiting a social sphere of civil but apolitical identity.

Failing to grasp the previous transformation from historical opt-out to hyperhistorical opt-in means being less likely to understand the apparent inconsistency between the disenchantment of individuals with politics and the popularity of global movements, international mobilisations, activism, voluntarism, and other social forces with huge political implications.[12] What is moribund is not politics *tout court*, but historical politics, that based on parties, classes, fixed social roles, political manifestos and programs, and the sovereign state, which sought political legitimacy only once and spent it until revoked. The inching towards the so-called centre by parties in liberal democracies around the world, as well as the 'get out the vote' strategies (the expression is used to describe the mobilisation of *voters as supporters* to ensure that those who can do vote) are evidence that engagement needs to be constantly renewed and expanded in order to win an election. Party (as well as Union) membership is a modern feature that is likely to become increasingly less common.

*Social vs. Political Space* In prehistory, the social and the political spaces overlap because, in a stateless society, there is no real difference between social and political relations and hence interactions. In history, the state tends to maintain such co-extensiveness by occupying, as an informational multiagent system, the social space politically, thus establishing the primacy of the political over the social. This trend, if unchecked and unbalanced, risks leading to totalitarianisms (consider for example the Italy of Mussolini), or at least broken democracies (consider next the Italy of Berlusconi). We have seen earlier that such co-extensiveness and its control may be based on normative or economic strategies, through the exercise of power, force, and rule-making. In hyperhistory, the social space is the original, default space from which agents may move to (consent to) join the political space. It is not accidental that concepts such as *civil society*,[13] *public sphere*,[14] and *community* become increasingly important the more we move into a hyperhistorical context. The problem is to understand and design such social space where agents of various kinds are supposed to be interacting and which give rise to the political multiagent system.

Each agent within the social space has some degrees of freedom. By this I do not mean liberty, autonomy or self-determination, but rather, in the robotic, more humble sense, some capacities or abilities, supported by the relevant resources, to engage in specific actions for a specific purpose. To use an elementary example, a coffee machine has only one degree of freedom: it can make coffee, once the right ingredients and energy are supplied. The sum of an agent's degrees of freedom is its 'agency'. When the agent is alone, there is of course only agency, but no social let alone political space. Imagine Robinson Crusoe on his 'Island of Despair'. However, as soon as there is another agent (Friday on the 'Island of Despair'), or indeed a group of agents (the native cannibals, the shipwrecked Spaniards, the English mutineers), agency acquires the further value of social interaction. Practices and then rules for coordination and constraint of the agents' degrees of freedom become

[12] On volunteerism see United Nations (2011). *State of the World's Volunteerism Report, 2011: Universal Values for Global Well-being*, United Nations Volunteers., on digital activism, the Digital Activism Research Project (http://digital-activism.org/) offers a wealth of information.

[13] I use the expression here in the post-Hegelian sense of non-political society.

[14] The social space where people can meet, identify and discuss societal problems, shaping political actions.

essential, initially for the well-being of the agents constituting the multiagent system, and then for the well-being of the multiagent system itself. Note the shift in the level of analysis: once the social space arises, we begin to consider the group as a group—e.g., as a family, or a community, or as a society—and the actions of the individual agents constituting it become elements that lead to the newly established degrees of freedom, or agency, of the multiagent system. The previous simple example may still help. Consider now a coffee machine and a timer: separately, they are two agents with different agency, but if they are properly joined and coordinated into a multiagent system, then the issuing agent has the new agency to make coffee at a set time. It is now the multiagent system that has a more complex capacity, and that may or may not work properly.

A social space is the totality of degrees of freedom of the inhabiting agents one wishes to take into consideration. In history, such consideration—which is really just another level of analysis—was largely determined physically and geographically, in terms of presence in a territory, and hence by a variety of forms of neighbourhood. In the previous example, all the agents interacting with Robinson Crusoe are taken into consideration because of their relations (interactive presence in terms of their degrees of freedom) to the same 'Island of Despair'. We saw that ICTs have changed all this. In hyperhistory, where to draw the line to include, or indeed exclude, the relevant agents whose degrees of freedom constitute the social space has become increasingly a matter of at least implicit choice, when not of explicit decision. The result is that the phenomenon of distributed morality, encompassing that of distributed responsibility, is becoming more and more common. In either case, history or hyperhistory, what counts as a social space may be a political move. Globalisation is a de-territorialisation in this political sense.

Turning now to the political space in which the new multiagent systems operate, it would be a mistake to consider it a separate space, over and above the social one. Both the social and the political space are determined by the same totality of the agents' degrees of freedom. The political space emerges when the complexity of the social space requires the prevention or resolution of potential *divergences* and coordination or collaboration about potential *convergences*. *Both* are crucial. And in each case information is required, in terms of representation and deliberation about a complex multitude of degrees of freedom.

*Legitimacy* It is when the agents in the social space agree to agree on how to deal with their divergences (conflicts) and convergences that the social space acquires the political dimension to which we are so used. Yet two potential mistakes await us here.

The first, call it Hobbesian, is to consider politics merely as the prevention of war by other means, to invert the famous phrase by von Clausewitz (1780–1831), according to whom 'war is the continuation of politics by other means'. This is an unsatisfactory view of politics, because even a complex society of angels would still require rules in order to further its harmony. Convergences too need politics. Politics is not just about conflicts due to the agents' exercises of their degrees of freedom when pursuing their goals. It is also, or at least it should be, above all, the furthering of coordination and collaboration of degrees of freedom by means other than coercion and violence.

The second potential mistake, which may be called Rousseauian, is to misunderstand the political space as just that part of the social space organised by law. In this case, the mistake is subtler. We usually associate the political space with the rules or laws that regulate it but the latter are not constitutive, by themselves, of the political space. Compare two cases in which rules determine a game. In chess, the rules do not merely constrain the game; they are the game because they do not supervene on a previous activity. Rather, they are the *necessary and sufficient conditions* that determine all and only the moves that can be legally made. In football, however, the rules are supervening *constraints* because the agents enjoy a previous and basic degree of freedom, consisting in their capacity to kick a ball with the foot in order to score a goal, which the rules are supposed to regulate. Whereas it is physically possible, but makes no sense, to place two pawns on the same square of a chessboard, nothing impeded Maradona from scoring an infamous goal by using his hand in the Argentina vs. England football match (1986 FIFA World Cup), and that to be allowed by a referee who did not see the infringement. Now, the political space is not simply *constituted* by the laws that regulate it, as in the chess example. But it is not just the result of the *constraining* of the social space by means of laws either, as in the football example. The political space is that area of the social space *configured* by the agreement to agree on resolution of divergences and coordination of convergences. The analogy here is the formatting of a hard disk. This leads to a further consideration, concerning the transparent multiagent system, especially when, in this transition time, the multiagent system in question is still the state.

*Transparency* There are two senses in which the multiagent system can be transparent. They mean quite different things, and so they can be confusing. Unsurprisingly, both come from ICTs and computer science, one more case in which the information revolution is changing our conceptual framework.

On the one hand, the multiagent system (think of the sovereign state, and also of corporate agents, multinationals, or supranational institutions, etc.) can be transparent in the sense that it moves from being a black box to being a white box. Other agents (citizens, when the multiagent system is the state) not only can see inputs and outputs—for example, levels of tax revenue and public expenditure—they can also monitor how (in our running example, the state as) a multiagent system works internally. This is not a novelty at all. It was a principle already popularised in the 19th century. However, it has become a renewed feature of contemporary politics due to the possibilities opened up by ICTs. This kind of transparency is also known as *Open Government*.

On the other hand, and this is the more innovative sense that I wish to stress here, the multiagent system can be transparent in the sense of being 'invisible'. This is the sense in which a technology (especially an interface) is transparent: not because it is not there, but because it delivers its services so efficiently, effectively, and reliably that its presence is imperceptible. When something works at its best, behind the scenes as it were, to make sure that we can operate as smoothly as possible, then we have a transparent system. When the multiagent system in question is the state, this second sense of transparency should not be seen as a surreptitious way of introducing, with a different terminology, the concept of 'small state' or 'small governance'. On the contrary, in this second sense, the multiagent system (the state) is as transparent and as vital as the oxygen that we breathe. It strives to be the ideal butler. There is no standard terminology for this kind of transparent multiagent system that becomes perceivable only when it is absent. Perhaps one may speak of *Gentle Government.*

It seems that multiagent systems can increasingly support the right sort of ethical infrastructure (more on this later) the more transparently, that is, openly and gently, they play the negotiating game through which they take care of the *res publica.* When this negotiating game fails, the possible outcome is an increasingly violent conflict among the parties involved. It is a tragic possibility that ICTs have seriously reshaped.

All this is not to say that *opacity* does not have its virtues. Care should be exercised, lest the socio-political discourse is reduced to the nuances of higher quantity, quality, intelligibility, and usability of information and ICTs. The more the better is not the only, nor always the best, rule of thumb. For the withdrawal of information can often make a positive and significant difference. We already encountered Montesquieu's division of the state's political powers. Each of them may be informationally opaque in the right way to the other two. For one may need to lack (or intentionally preclude oneself from accessing) some information in order to achieve desirable goals, such as protecting anonymity, enhancing fair treatment, or implementing unbiased evaluation. Famously, Rawls' 'veil of ignorance' exploits precisely this aspect of information, in order to develop an impartial approach to justice.[15] Being informed is not always a blessing and might even be dangerous or wrong, distracting or crippling. The point about the value of transparency is that its opposite, informational opacity, cannot be assumed to be a good property of a political system unless it is adopted explicitly and consciously, by showing that it is a feature not a mere bug.

### *6.1.4 Infraethics*

Part of the ethical efforts engendered by the fourth revolution concerns the design of environments that can facilitate ethical choices, actions, or process. This is not the same as *ethics by design.* It is rather *pro-ethical design*, as I hope will become clearer in the following pages. Both are liberal, but *ethics by design* may be mildly paternalistic, insofar as it privileges the facilitation of the *right* kind of choices, actions, process or interactions on behalf of the agents involved. Whereas *pro-ethical design* does not have to be paternalistic, insofar as it privileges the facilitation of *reflection* by the agents involved on their choices, actions, or process. For example,

[15] Rawls (1999).

strategies based on *ethics by design* may let you opt out of the *default* preference according to which, by obtaining a driving licence, you are also willing to be an organ donor. Strategies based on *pro-ethical design* may not allow you to obtain a driving license unless you have indicated whether you wish to be an organ donor, the unbiased choice is still all yours. In this section, I shall call environments that can facilitate ethical choices, actions, or process, the ethical infrastructure, or *infraethics*. The problem is how to design the right sort of infraethics. Clearly, in different cases, the design of a liberal infraethics may be more or less paternalistic. My argument is that it should be as little paternalistic as the circumstances permit, although no less.

It is a sign of the times that, when politicians speak of infrastructure nowadays, they often have in mind ICTs. They are not wrong. From business fortunes to conflicts, what makes contemporary societies work depends increasingly on bits rather than atoms. We already saw all this. What is less obvious, and intellectually more interesting, is that ICTs seem to have unveiled a new sort of ethical equation.

Consider the unprecedented emphasis that ICTs have placed on crucial phenomena such as trust, privacy, transparency, freedom of expression, openness, intellectual property rights, loyalty, respect, reliability, reputation, rule of law, and so forth. These are probably better understood in terms of an infrastructure that is there to facilitate or hinder (reflection upon) the im/moral behaviour of the agents involved. Thus, by placing our informational interactions at the centre of our lives, ICTs seem to have uncovered something that, of course, has always been there, but less visibly so: the fact that the moral behaviour of a society of agents is also a matter of 'ethical infrastructure' or simply *infraethics*. An important aspect of our moral lives has escaped much of our attention. Many concepts and related phenomena have been mistakenly treated as if they were only ethical, when in fact they are probably mostly infraethical. To use a term from the philosophy of technology, such concepts and the corresponding phenomena have a dual-use nature: they can be morally good, but also morally evil (more on this presently). The new equation indicates that, in the same way that, in an economically mature society, business and administration systems increasingly require infrastructures (transport, communication, services etc.) to prosper, so too, in an informationally mature society, multiagent systems' moral interactions increasingly require an infraethics to flourish.

The idea of an infraethics is simple, but can be misleading. The previous equation helps to clarify it. When economists and political scientists speak of a 'failed state', they may refer to the failure of a *state-as-a-structure* to fulfil its basic roles, such as exercising control over its borders, collecting taxes, enforcing laws, administering justice, providing schooling, and so forth. In other words, the state fails to provide *public goods*, such as defence and police, and *merit goods*, such as healthcare. Or (too often an inclusive and intertwined or) they may refer to the collapse of a *state-as-an-infrastructure* or environment, which makes possible and fosters the right sort of social interactions. This means that they may be referring to the collapse of a substratum of default expectations about economic, political and social conditions, such as the rule of law, respect for civil rights, a sense of political community, civilised dialogue among differently-minded people, ways to reach peaceful resolutions of ethnic, religious, or cultural tensions, and so forth. All these expectations, attitudes, practices, in short such an implicit 'socio-political infrastructure', which one may take for granted, provides a vital ingredient for the success of any complex society. It plays a crucial role in human interactions, comparable to the one that we are now accustomed to attributing to physical infrastructures in economics.

Infraethics should not be understood in terms of Marxist theory, as if it were a mere update of the old 'base and superstructure' idea. The elements in question are entirely different: we are dealing with moral actions and not-yet-moral facilitators of such moral actions. Nor should it be understood in terms of a kind of second-order normative discourse on ethics. It is the not-yet-ethical framework of implicit expectations, attitudes, and practices that *can* facilitate and promote moral decisions and actions. At the same time, it would also be wrong to think that an infraethics is morally neutral. Rather, it has a dual-use nature, as I anticipated earlier: it can both facilitate and hinder morally good as well as evil actions, and do this in different degrees. At its best, it is the grease that lubricates the moral mechanism. This is more likely to happen whenever having a 'dual-use' nature does not mean that each use is equally likely, that is, that the infraethics in question is still not neutral, nor merely positive, but does have a bias to deliver more good than evil. If this is confusing, think of the dual-use nature not in terms of a state of equilibrium, like an ideal coin that can deliver both heads and tails, but in terms of a co-presence of two alternative outcomes, one of which is more likely than the other, as a biased coin more likely to turn heads than tails. When an infraethics has a 'biased dual-use' nature, it is easy to mistake the infraethical for the ethical, since whatever helps goodness to flourish or evil to take root partakes of their nature.

Any successful complex society, be this the City of Man or the City of God, relies on an implicit infraethics. This is dangerous, because the increasing importance of an infraethics may lead to the following risk: that the legitimization of the ethical discourse is based on the 'value' of the infraethics that is supposed to support it. *Supporting* is mistaken for *grounding*, and may even aspire to the role of *legitimizing*, leading to what the French philosopher Jean-François Lyotard (1924–1998) criticized as mere 'performativity' of the system, independently of the actual values cherished and pursued. As an example, think of a bureaucratic context in which some procedure, supposed to deliver some morally good behavior, through time becomes a value in itself, and ends giving ethical value to the behavior that was supposed to support. Infraethics is the vital syntax of a society, but it is not its semantics, to re-use a distinction we encountered when discussing artificial intelligence. It is about the structural form, not the meaningful contents.

We saw earlier that even a society in which the entire population consisted of angels, that is, perfectly moral agents, still needs norms for collaboration and coordination. Theoretically, a society may exist in which the entire population consisted of Nazi fanatics who could rely on high levels of trust, respect, reliability, loyalty, privacy, transparency, and even freedom of expression, openness, and fair competition. Clearly, what we want is not just the successful mechanism provided by the right infraethics, but also the coherent combination between it and morally good values, such as civil and political rights. This is why a balance between security and privacy, for example, is so difficult to achieve, unless we clarify first whether we are dealing with a tension within ethics (security and privacy as moral rights), within infraethics (both are understood as not-yet-ethical facilitators), or between infraethics (security) and ethics (privacy), as I suspect. To rely on another analogy: the best pipes (infraethics) may improve the flow but do not improve the quality of the water (ethics); and water of the highest quality is wasted if the pipes are rusty or leaky. So creating the right sort of infraethics and maintaining it is one of the crucial challenges of our time, because an infraethics is not morally good in itself, but it is what is most likely to yield moral goodness if properly designed and combined with the right moral values. The right sort of infraethics should be there to support the right sort of values. It is certainly a constitutive part of the problem concerning the design of the right multiagent systems.

The more complex a society becomes, the more important and hence salient the role of a well-designed infraethics is, and yet this is exactly what we seem to be missing. Consider the recent Anti-Counterfeiting Trade Agreement (ACTA), a multinational treaty concerning the international standards for intellectual property rights.[16] By focusing on the enforcement of intellectual property rights (IPR), supporters of ACTA completely failed to perceive that it would have undermined the very infraethics that they hoped to foster, namely one promoting some of the best and most successful aspects of our information society. It would have promoted the structural inhibition of some of the most important individuals' positive liberties and their ability to participate in the information society, thus fulfilling their own potential as informational organisms. For lack of a better word, ACTA would have promoted a form of *informism*, comparable to other forms of social agency's inhibition such as classism, racism, and sexism. Sometimes a protection of liberalism may be inadvertently illiberal. If we want to do better, we need to grasp that issues such as IPR are part of the new infraethics for the information society, that their protection needs to find its carefully balanced place within a complex legal and ethical infrastructure that is already in place and constantly evolving, and that such a system must be put at the service of the right values and moral behaviours. This means finding a compromise, at the level of a liberal infraethics, between those who see new legislation (such as ACTA) as a simple fulfilment of existing ethical and legal obligations (in this case from trade agreements), and those who see it as a fundamental erosion of existing ethical and legal civil liberties.

In hyperhistorical societies, any regulation affecting how people deal with information is now bound to influence the whole infosphere and onlife habitat within which they live. So enforcing rights such as IPR becomes an environmental problem. This does not mean that any legislation is necessarily negative. The lesson here is one about complexity: since rights such as IPR are part of our infraethics and affect our whole environment understood as the infosphere, the intended and unintended consequences of their enforcement are widespread, interrelated, and far-reaching. These consequences need to be carefully considered, because mistakes will generate huge problems that will have cascading costs for future generations, both ethically and economically. The best way to deal with 'known unknowns' and

[16] For a more detailed analysis see Floridi (2012).

unintended consequences is to be careful, stay alert, monitor the development of the actions undertaken, and be ready to revise one's decision and strategy quickly, as soon as the wrong sort of effects start appearing. *Festina lente*, 'more haste, less speed' as the classic adage suggests. There is no perfect legislation but only legislation that can be perfected more or less easily. Good agreements about how to shape our infraethics should include clauses about their timely updating.

Finally, it is a mistake to think that we are like outsiders ruling over an environment different from the one we inhabit. Legal documents (such as ACTA) emerge from within the infosphere that they affect. We are building, restoring, and refurbishing the house from inside. Recall that we are repairing the raft while navigating on it, to use the metaphor introduced in the Preface. Precisely because the whole problem of respect, infringement, and enforcement of rights such as IPR is an infraethical and environmental problem for advanced information societies, the best thing we can do, in order to devise the right solution, is to apply to the process itself the very infraethical framework and ethical values that we would like to see promoted by it. This means that the infosphere should regulate itself from within, not from an impossible without.

### *6.1.5 Hyperhistorical Conflicts and Cyberwar*

The story goes that when the Roman horsemen first saw Pyrrhus' twenty war elephants, at the battle of Heraclea (280 BC), they were so terrorised by these strange creatures, which they had never seen before, that they galloped away, and the Roman legions lost the battle. Today, the new elephants are digital. The phenomenon might have just begun to emerge in the public debate but, in hyperhistorical societies, ICTs are increasingly shaping armed conflicts.

Disputes become armed conflicts when politics fails. In hyperhistory, such armed conflicts have acquired a new informational nature. Cyberwar or information warfare is the continuation, and sometimes the replacement, of conflict by digital means, to rely once more on von Clausewitz's famous interpretation of war we encountered above. Four main changes are notable.

First, in terms of conventional military operations, ICTs have progressively revolutionized communications, making possible complex new modes of field operations. We saw this was already the case with the *Chappe* telegraph.

Second, ICTs have also made possible the swift analysis of vast amounts of data, enabling the military, intelligence and law enforcement communities to take action in ever more timely and targeted ways. ICTs and Big Data are also weapons.

Third, and even more significantly, battles are nowadays fought by highly mobile forces, armed with real-time ICT devices, satellites, battlefield sensors and so forth, as well as thousands of robots of all kinds.

And, finally, the growing dependence of societies and their militaries on advanced ICTs has led to strategic cyber-attacks, designed to cause costly and crippling disruption. Armies of human soldiers may no longer be needed. This creates a stark contrast with suicide terrorism. On the one hand, human life can regain its ultimate value because the state no longer needs to trump it in favour of patriotism. Contrary to what we saw in the previous pages, drones do not die 'for King and Country'. Cyberwar is a hyperhistorical phenomenon. On the other hand, terrorists de-humanise individuals as mere delivery mechanisms. Suicide terrorism is a historical phenomenon, in which the technology in-between is the human body and a person becomes a 'living tool', using Aristotle's definition of a slave.

The old economic problem—how to finance war and its expensive high tech—is now joined by a new legal problem: how to reconcile a hyperhistorical kind of warfare with historical phenomena, such as the infringement of national sovereignty and respect for geographical borders. Furthermore, cyber-attacks can be undertaken by nations or networks, or even by small groups or individuals. ICTs have made asymmetric conflicts easier, and shifted the battleground more than an inch into the infosphere.

The scale of such transformations is staggering. For example, in 2003, at the beginning of the war in Iraq, US forces had no robotic systems on the ground. However, by 2004, they had already deployed 150 robots, in 2005 the number was 2,400; and by the end of 2008, about 12,000 robots of nearly two dozen varieties were operating on the ground.[17]

In 2010, Neelie Kroes, Vice-President of the European Commission, commenting on Cyber Europe 2010, the first pan-European cyber-attack simulation, said that:

> This exercise to test Europe's preparedness against cyber threats is an important first step towards working together to combat potential online threats to essential infrastructure and ensuring citizens and businesses feel safe and secure online.[18]

As you can see, the perspective could not be more hyperhistorical.

ICT-mediated modes of conflict pose a variety of ethical problems, for war-fighting militaries in the field, for intelligence gathering services, for policy makers, and for ethicists. They may be summarised as the three Rs: risks, rights and responsibilities.

*Risks* Cyberwar and information-based conflicts may increase risks, making 'soft' conflicts more likely and hence potentially increasing the number of casualties. Between 2004 and 2012, drones operated by the US' Central Intelligence Agency (CIA) killed more than 2,400 people in Pakistan, including 479 civilians, with 3 strikes in 2005 escalating to 76 strikes in 2011.[19] A troubling perspective is that ICTs might make unconventional conflicts more acceptable ethically, by stressing the less deadly outcome of military operations in cyberspace. However, this might be utterly illusory. Messing with ICT-infrastructures of hospitals and airports may easily cause the loss of human lives, even if in a less obvious way than bombs do.

---

[17] Source: *The New Atlantis* report, available online.

[18] Source: Press release, *Digital Agenda: cyber-security experts test defences in first pan-European simulation*, available online.

[19] The Economist (2012).

Despite this, the mistaken impression remains that we might be allegedly moving towards a more precise, surgical, bloodless way of handling violently our political disagreements.

*Rights* Cyberwar tends to erase the threshold between reality and simulation, between life and play, and between conventional conflicts, insurgencies or terrorist actions. This threatens to increase the potential tensions between fundamental rights: informational threats require higher levels of control, which may generate conflicts between individuals' rights (e.g. privacy) and community's rights (e.g. safety and security). A state's duty to protect its citizens may come to clash with its duty to prevent harm to its citizens, via an extended system of surveillance, which may easily end up infringing on citizens' privacy.

*Responsibilities* Cyberwar makes it more difficult to identify responsibilities that are reshaped and distributed. Because causal links are much less easily identifiable, it becomes much more difficult to establish who, or what, is accountable and responsible when software/robotic weapons and hybrid, man-machine systems are involved.

New risks, rights and responsibilities: in short, cyberwar is a new phenomenon, which has caught us by surprise. With hindsight, we should have known better, for at least three reasons.

Take the nature of our society first. When it was modern and industrial, conflicts had mechanised, second-order features. Engines, from battleships to tanks to aeroplanes, were weapons, and the coherent outcome was the emphasis on energy, petrol first and then nuclear power. There was an eerie analogy between assembly lines and warfare trenches, between working force and fighting force. Conventional warfare was kinetic warfare. We just did not know it, because the non-kinetic kind was not yet available. The Cold War and the emergence of asymmetric conflicts were part of a post-industrial transformation. Today, in a culture in which we have seen that the word 'engine' is more likely to be preceded by the verb 'search' than by the noun 'petrol', hyperhistorical societies are as likely to fight with digits as they are with bullets, with computers as well as guns, not least because digital systems tend to be in charge of analogue weapons. I am not referring to the use of intelligence, espionage, or cryptography, but to cyber attacks or to the extensive use of drones and other military robots in Iraq and Afghanistan. It is old news. On 27th of April 2007, about 1 million computers worldwide were used for DDOS (distributed denial of service) attacks on Estonian government and corporate web sites. A DDOS attack is a systematic attempt to make computer resources unavailable, at least temporarily, by forcing vital sites or services to reset or consume their resources, or by disrupting their communications so that they can no longer function properly. Russia was blamed but denied any involvement. In June 2010, Stuxnet, a sophisticated computer malware, sabotaged ca. 1000 Siemens centrifuges used in the Iranian nuclear power plant of Bushehr. That time, the US and Israel denied any involvement. At the time of writing, there is an on-going attack on US ICT infrastructure. This time China that denies any involvement. Then there are robotic weapons, which may be seen as the final stage in the industrialisation of warfare, or, more interestingly, as the first step in the development of information conflicts, in which command and control as well as action and reaction become tele-concepts. Third-order technological conflicts in which humans are no longer in the loop have moved out of science fiction and into military scenarios. From software agents in cyberspace to robots in physical environments we should not be too optimistic about the non-violent nature of cyberwar. The more we rely on ICTs, the more we envelop the world, the more cyber attacks will become lethal. Soon, crippling an enemy's communication and information infrastructure will be like zapping its pacemaker rather than hacking its mobile.

Second, consider the nature of our environment. We have been talking about the internet and cyberspace for decades. We could have easily imagined that this would become the new frontier for human conflicts. Technologies have continuously expanded. We have been fighting each other on land, at sea, in the air, and in space for as long, and as soon as technologies made it possible. Predictably, the infosphere was never going to be an exception. Information is the fifth element,[20] and the military now speaks of cyberwarfare as 'the fifth domain of warfare'. The impression is that, in the future, such a fifth domain will end up dominating the others. The following two examples may help. On 13th of May 1999, arguably the first combat between an aircraft and an unmanned drone took place when an Iraqi MiG-25 shot down a US Air Force unmanned MQ-1 Predator drone. More than 360 drones have been built since 1995, for more than $ 2.38 billion. Second, since 2006, Samsung, the maker of smart phones and refrigerators, has also been producing the SGR-A1. It is a robot with a low-light camera and pattern recognition software to distinguish humans from animals or other objects. It patrols South Korea's border with North Korea and, if necessary, it can autonomously fire its built-in machine gun. It is increasingly hard to draw a clear distinction between cyberwarfare and conventional, kinetic warfare when some tele-warfare is in question.

Finally, think of the origin of cybernetics, the computer, the Internet, the Global Positioning System (GPS), and unmanned drones and vehicles. They all developed initially as part of wider military efforts. The history of computing is deeply rooted in the Second World War and Turing's work at Bletchley Park. Cybernetics, the ancestor of contemporary robotics, begun to develop as an engineering field in connection with applications for the automatic control of gun mounts and radar antenna, still during the Second World War. We know that the internet was the outcome of the arms race and of nuclear proliferation, but we were distracted by the development of the Web and its scientific origins, and forgot about the Defense Advanced Research Projects Agency (DARPA). The now ubiquitous GPS, which provides the satellite-based information for navigation systems, was created and developed by the US Department of Defence, one more case of the political importance of geography. It became freely available for civilian use only in 1983, after a Boeing 747 of the Korean Air Lines, with 269 people on board, was shot down because it had strayed into the USSR's prohibited airspace. Finally, the development of drones, mainly but not only by the US military, as well as autonomous vehicles (DARPA

[20] Floridi (1999).

again) and other robots, owes much to the conflicts in Iraq and Afghanistan and the fight against terrorism. In short, much of the history of digital ICTs spookily corresponds to the history of conflicts and the financial efforts behind them: Second World War, Cold War, First and Second Iraq War, War in Afghanistan, and various 'wars' on terrorist organisations around the world. Hyperhistory has merely caught up with us.

The previous outline should help one understand why cyberwar, or more generally information warfare, is causing radical transformations in our ways of thinking about military, political, and ethical issues. The concepts of state, war, and the distinction between civil society and military organisations are being affected. Are we going to see a new arms race, given the high rate at which cyber weapons "decay"? After all, you can use a piece of malware only once, for a patch will then become available, and often only within, and against, a specific technology that will soon be out of date. If cyber disarmament is ever going to be an option, how do you decommission cyber weapons? Digital systems can be hacked: will the Pony Express make a patriotic comeback in the near future as the last line of defence against an enemy that could tamper with anything digital and online? Some questions make one smile, but others are increasingly problematic. Let me highlight two sets of them that should be of more general interest.

The body of knowledge and discussion behind Just War Theory is detailed and extensive.[21] It is the result of centuries of refinements since Roman times. The methodological question we face today is whether information warfare is merely one more area of application, or whether it represents a disruptive novelty as well, which will require new developments of the theory itself. For example, within the *jus ad bellum*, which kind of authorities possesses the legitimacy to wage cyberwar? And how should a cyber attack be considered in terms of last resort, especially when a cyber attack could, allegedly, prevent more violent outcomes? And within the *jus in bello*, what level of proportionality should be attributed to a cyber attack? How do you surrender to cyber enemies, especially when their identities are unknown on purpose? Or how will robots deal with non-combatants or treat prisoners? Is it possible or even desirable to develop in-built 'ethical algorithms' when engineering robotic weapons?

Equally developed, in this case since Greek times, is our understanding of military virtue ethics. How is the latter going to be applied to phenomena that are actually reshaping the conditions of possibility of virtue ethics itself? Bear in mind that any virtue ethics presupposes a philosophical anthropology, that is, a view of the human nature that may be Aristotelian, Buddhist, Christian, Confucian, Fascist, Nietzschean, Spartan, and so forth. Information warfare is only part of the information revolution, which is also affecting our self-understanding as informational organisms. Take for example the classic virtue of courage: in what sense can someone be courageous when tele-manoeuvring a military robot? Indeed, will courage still rank so highly among the virtues when the capacity to evaluate and manage information

[21] For a study of how current international law applies to cyber conflicts and cyber warfare, see NATO Cooperative Cyber Defence Centre of Excellence (2013).

and act upon it wisely and promptly will seem to be a much more important trait of a soldier's character?

Similar questions seem to invite new theorising, rather than the mere application or adaptation of old ideas. ICTs have caused radical changes both in how societies may come into conflict and how they may manage it. At the same time, there is a policy and a conceptual deficit. For example, the US Department of Defence intends to replace a third of its armed vehicles and weaponry with robots by 2015, but it still lacks an ethical code for the deployment of these new, semi-autonomous weapons.[22] This is a global issue. The 2002 Prague Summit marked NATO's first attempt to address cyber-defence activities. Five years later, in 2007, there were already 42 countries working on military robotics, including Iran, China, Belarus, and Pakistan,[23] but not even a draft of an international agreement regarding their ethical deployment. There is a serious need for more descriptive and conceptual analyses of such a crucial area in applied ethics, and more assessment of the effectiveness of the initial measures that have been taken to deal with the increasing application of ICTs in armed conflicts. The issue could not be more pressing and there is a much felt and quickly escalating need to share information and coordinate ethical theorising. The goals should be sharing information and views about the current state of the ethics of information warfare, developing a comprehensive framework for a clear interpretation of the new aspects of cyberwar, building a critical consensus about the ethical deployment of e-weapons, and laying down the foundation for an ethical approach to information warfare. We experimented with chemical weapons, especially during the First World War, and with biological weapons, in particular during the Sino-Japanese War of 1931–1945. The horrific results led, in 1925, to the Geneva Protocol, prohibiting the use of chemical and biological weapons. In 1972, the Biological and Toxin Weapons Convention (BWC) banned the development, production and storage of bio-weapons. Since then, we have managed to restrain their use and, by and large, respect the BWC. Something similar happened with nuclear weapons. The hope is that information warfare and e-weapons will soon be equally regulated and constrained, without having to undergo any terrible and tragic lesson.

Let us return to the elephants. During the civil war, in the battle of Thapsus (46 BC), Julius Caesar's fifth legion was armed with axes and was ordered to strike at the legs of the enemy's elephants. The legion withstood the charge, and the elephant became its symbol. Interestingly, nobody at the time could even imagine that there might be an ethical problem in treating animals so cruelly. We should think ahead, because history occasionally is a bit petulant and likes to repeat itself. At a time when there is an exponential growth in R&D concerning ICT-based weapons and strategies, we should collaborate on the identification, discussion and resolution of the unprecedented ethical difficulties characterizing cyberwar. This is far from being premature. Perhaps, instead of updating our old ethical theories with more and more service packs, we might want to consider upgrading them by developing new ideas. Like the civilian uses of robots, information warfare calls for an information

[22] The Economist (2007).

[23] Source: *The Wilson Quarterly*, report available online.

ethics. After all, iRobot produces both the *Roomba 700* that vacuum cleans your floor and the *iRobot 710 Warrior* that disposes of your enemies' explosives.

### 6.1.6 Conclusion

Six thousand years ago, humanity witnessed the invention of writing and the emergence of the conditions of possibility that were going to lead to cities, kingdoms, empires, sovereign states, nations, and intergovernmental organisations. This is not accidental. Prehistoric societies are both ICT-less and stateless. The state is a typical historical phenomenon. It emerges when human groups stop living a hand-to-mouth existence in small communities and begin to live a mouth-to-hand. Large communities become political societies, with division of labour and specialised roles, organised under some form of government, which manages resources through the control of ICTs, including that special kind of information called 'money'. From taxes to legislation, from the administration of justice to military force, from census to social infrastructure, the state was for a long time the ultimate information agent and so history, and especially modernity, is the age of the state.

Almost halfway between the beginning of history and now, Plato was still trying to make sense of both radical changes: the encoding of memories through written symbols and the symbiotic interactions between the individual and the *polis*–state. In 50 years, our grandchildren may look at us as the last of the historical, state-organised generations, not so differently from the way we look at some Amazonian tribes, as the last of the prehistorical, stateless societies. It may take a long while before we come to understand in full such transformations.

## References

Anderson, S., and J. Cavanagh. 2000. Top 200: The rise of corporate global power. Institute for Policy Studies 4.

Floridi, L. 1999. *Philosophy and computing: An introduction*. London: Routledge.

Floridi, L. 2012. ACTA—The ethical analysis of a failure, and its lessons. ECIPE working papers 04/2012.

Halper, S. A. 2010. *The Beijing consensus: How China's authoritarian model will dominate the twenty-first century*. New York: Basic Books.

NATO Cooperative Cyber Defence Centre of Excellence. 2013. *Tallinn manual on the international law applicable to cyber warfare: prepared by the international group of experts at the invitation of the NATO Cooperative Cyber Defence Centre of Excellence*. Cambridge: Cambridge University Press.

Ramo, J. C., and Foreign Policy Centre (London England). 2004. *The Beijing consensus*. London: Foreign Policy Centre.

Rawls, J. 1999. *A theory of justice*. Cambridge: Belknap Press of Harvard University Press.

The Economist. June 7 2007. Robot wars.

The Economist. June 2 2012. Morals and the machine.

United Nations. 2011. *State of the world's volunteerism report, 2011: Universal values for global well-being*. United Nations volunteers.
Williamson, J. 1993. Democracy and the "Washington consensus". *World Development* 21 (8): 1329–1336.
Williamson, J. 2012. Is the "Beijing Consensus" now dominant? *Asia Policy* 13 (1): 1–16.

# Chapter 7
# An Ethical Framework for Information Warfare

**Mariarosaria Taddeo**

**Abstract** In this Chapter I propose an ethical analysis of information warfare, the warfare waged in the cyber domain. The goal is twofold, filling the theoretical vacuum surrounding this phenomenon and providing the conceptual grounding for the definition of new ethical regulations for information warfare. I argue that Just War Theory is a necessary but not sufficient instrument for considering the ethical implications of information warfare and that a suitable ethical analysis of this kind of warfare is developed when Just War Theory is merged with Information Ethics. In the initial part of the chapter, I describe information warfare and its main features and highlight the problems that arise when Just War Theory is endorsed as a means of addressing ethical problems engendered by this kind of warfare. In the final part, I introduce the main aspects of Information Ethics and define three principles for a just information warfare resulting from the integration of Just War Theory and Information Ethics.

## 7.1 Introduction

Since 2010, the cyberspace has been officially listed among the domains in which war may be waged these days. It comes fifth along land, sea, air and space, for the ability to control, disrupt or manipulate the enemy's informational infrastructure has become as decisive with respect to the outcome of conflicts as weapon superiority. Information and communication technologies (ICTs) have proved to be a useful and convenient technology for waging war, the military deployment of ICTs has radically changed the way wars are declared and waged nowadays. It has actually determined the latest revolution in military affairs, i.e. the informational

M. Taddeo (✉)
Political & International Studies, University of Warwick, Gibbet Hill Road,
Coventry CV4 7AL, UK
e-mail: m.taddeo@warwick.ac.u

University of Oxford, Oxford, UK

L. Floridi (ed.), *Protection of Information and the Right to Privacy – A New Equilibrium?*,
Law, Governance and Technology Series 17, DOI 10.1007/978-3-319-05720-0_7

turn in military affairs (Toffler and Toffler 1997; Taddeo 2012).[1] Such a revolution is not the exclusive concern of the military; it has also a bearing on ethicists and policymakers, since existing ethical theories of war and national and international regulations struggle to address the novelties of this phenomenon.

In this chapter I propose an ethical analysis of information warfare (IW) with the twofold goal of filling the theoretical vacuum surrounding this phenomenon and of providing the conceptual grounding for the definition of new ethical regulations for IW. The proposed analysis rests on the conceptual investigation of IW that I provided in (Taddeo 2012), where I highlight the informational nature of this phenomenon and maintain that IW represents a profound novelty, which is reshaping the very concept of war and raises the need for new ethical guidelines.

Following on that analysis, in this chapter I argue that considering IW through the lens of Just War Theory (JWT) allows for the unveiling of fundamental ethical issues that this phenomenon brings to the fore, yet that attempting to address these issues solely on the basis of this theory will leave them unsolved. I then suggest that problems encountered when addressing IW through JWT are overcome if the latter is merged with Information Ethics (Floridi 2013). This is a macro-ethical theory, which is particularly suitable for taking into account the features and the ethical implications of *informational phenomena*, like for example internet neutrality (Turilli et al. 2011), online trust, peer-to-peer (Taddeo and Vaccaro 2011) and IW. Merging the principles of JWT with the macro-ethical framework provided by Information Ethics has two advantages: it allows the development of an ethical analysis of IW capable of taking into account the peculiarities and the novelty of this phenomenon; it also extends the validity of JWT to a new kind of warfare, which at first glance seemed to fall outside its scope (Taddeo 2012).

In the initial part of this chapter, I describe IW and its main features, I will then focus on JWT and on the problems that arise when this theory is endorsed as a means of addressing the case for IW. Information Ethics will then be introduced, its four principles will provide the grounds for the analysis proposed in the final part of this chapter, where I describe the principles for a just IW and discuss how JWT can be applied to IW without leading to ethical conundrums. Having delineated the path ahead of us, we should now begin our analysis by considering in more detail the nature of IW.

## 7.2 Information Warfare

The expression 'information warfare' has already been used in the extant literature to refer solely to the uses of ICTs devoted to breaching the opponent's informational infrastructure in order to either disrupt it or acquire relevant data and information

[1] For an analysis of revolution in military affairs considering both the history of such revolutions and the effects of the development of the most recent technologies on warfare see (Benbow 2004; Blackmore 2011).

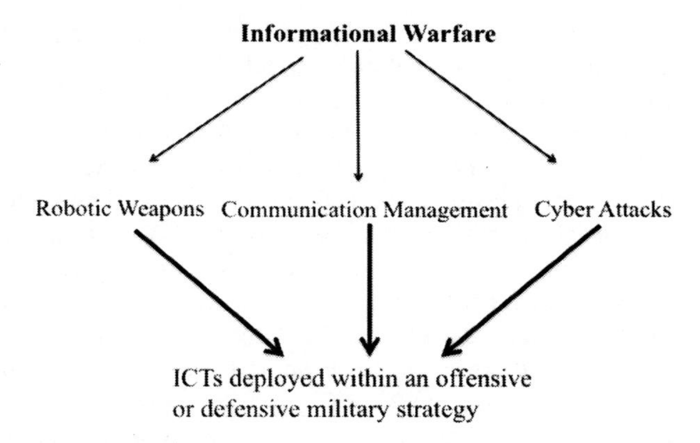


**Fig. 7.1** The different uses of ICTs in military strategies. (Taddeo 2012, p. 110)

about the opponent's resources, military strategies and so on; see for example (Libicki 1996; Waltz 1998; Schwartau 1994).

Distributed denials of service (DDoS) attacks, like the ones launched in Burma during the 2010 elections[2], the injection of Stuxnet in the Iranian nuclear facilities of Bushehr[3], as well as 'Red October' discovered in 2013 are all famous examples of how ICTs can be used to conduct the so-called cyber attacks.[4] Nonetheless, such attacks are only one of the instances of IW. In the rest of this chapter, I will use IW to refer to a wide spectrum of phenomena, encompassing cyber-attacks as well as the deployment of robotic-weapons and ICT-based communication protocols (see Fig. 7.1).

Endorsing a wide spectrum definition of IW offers important advantages, both conceptual and methodological. The conceptual advantage revolves around the identification of the informational nature of this phenomenon. In all the three cases, information plays a crucial role, it is either the target, the source or the medium for the accomplishment of a given goal. Now, while this is evident for the cases of communication management and cyber attacks, further explanation may be needed to highlight the informational nature of the deployment of (semi)autonomous robotic weapons, which may be less obvious. Such weapons are usually deployed (or designed to be deployed) to participate in traditional military actions and usually have destructive purposes. See for example Israel's Harpi[5] or Taranis.[6]

Nonetheless, while (semi) autonomous weapon may be used to perform tasks and achieve goals not dissimilar to the ones pursued in traditional warfare, their modes of operations are quite different from traditional ones as they rely extensively on

[2] http://www.bbc.co.uk/news/technology-11693214 http://news.bbc.co.uk/2/hi/europe/6665145.stm.

[3] http://www.cbsnews.com/stories/2010/11/29/world/main7100197.shtml.

[4] For an annotated time line of cyber attacks see NATO's website http://www.nato.int/docu/review/2013/Cyber/timeline/EN/index.htm.

[5] This is an autonomous weapon system designed to detect and destroy radar emitters http://www.israeli-weapons.com/weapons/aircraft/uav/harpy/harpy.html.

[6] This is a UK drone which can autonomously search, identify and locate enemies although it should be stressed that it can only engage with a target upon the authorization of mission command http://en.wikipedia.org/wiki/BAE_Systems_Taranis.

the collection and elaboration of information. The ability and the way in which a machine collects, manipulates and checks information against the requirements for an action to be performed are crucial for the accomplishment of the given task. Information is in this case the means for the achievement of the goal and it shows to be common aspect to all these three cases. In the rest of this chapter I endorse an informational level of abstraction (LoA) to focus on such a common factor.

A brief digression from the analysis of IW is in order here to introduce LoAs. Any given systems, for example a car, can be observed focusing on certain aspects and disregarding others, the choice of the aspects on which one focuses, i.e. the observables, depends on the purpose of the observer. An engineer interested in maximising the aerodynamics would focus on the shape of car's parts, their weight and possibly the materials of which the parts are made. A costumer interested in the aesthetics of the car will focus on its colour and on the overall look of the car. The engineer and the costumer observe the car endorsing different LoAs. A LoA is a finite but non-empty set of observables accompanied by a statement of what feature of the system under consideration such a LoA stands for. A collection of LoAs constitutes an interface. An interface is used when analysing a system from various points of view, that is, at varying LoAs. It is important to stress that a single LoA does not reduce a car to merely the aerodynamics of its parts or to its overall look. A LoA is a tool that helps to make explicit the observation perspective and constrain it to only those elements that are relevant in a particular observation.[7]

Endorsing an informational LoA to analyse cyber attacks, the deployment of robotic-weapons and ICT-based communication protocols allows for unveiling the common factor to these three phenomenon rather than their differences. A different (lower) LoA can be endorsed in a second moment in order to analyse the specific occurrences of these three phenomena and their ethical implications. This approach neither undermines the differences between the use of a computer virus, ICT-based communication protocols and robotic weapons nor denies that such different uses generate different ethical issues. Rather, it aims at focusing first on the aspects that are common among the military uses of ICTs, since the analysis of these aspects provides the groundwork for addressing specific ethical problems brought to the fore by the different modes of military deployment of ICTs.

The methodological advantage of endorsing a wide spectrum definition concerns the scope of the analysis, by considering indiscriminately the different uses of ICTs in warfare, the analysis will address the totality of the cases of IW rather than focusing solely on some of its specific occurrences.

IW is thus defined as follows:

> **Information Warfare** is the use of ICTs within an offensive or defensive military strategy endorsed by a [political authority] and aiming at the immediate disruption or control of the enemy's resources, and which is waged within the informational environment, with agents and targets ranging both on the physical and non-physical domains and whose level of violence may vary upon circumstances. (Taddeo 2012)

[7] For more in details analysis of LoA see (Floridi 2008).

This definition highlights two important aspects of IW, its *informational nature* and its *transversality*, which put it in relation with the so-called information revolution (Floridi 2014; Taddeo 2012). The information revolution is a complex phenomenon. It rests on the development and the ubiquitous dissemination of the use of ICTs, which have a wide impact on many of our daily practises: from our social and professional lives to our interactions with the environment surrounding us. ICTs allow for developing and acting in a new domain, the digital or informational one. This is a virtual, non-physical domain, which has grown important and hosts a considerable relevant part of our lives. With the information revolution we witness a shift, which has brought the *non-physical domain* to the fore and made it as important and valuable as the physical one (Taddeo 2012).

IW is one of the most compelling instances of such a shift. It shows that there is a new environment, where physical and non-physical entities coexist and are equally valuable, and in which states have to prove their authority and new modes of warfare are being specifically developed.[8] The shift toward the non-physical domain provides the ground for the transversality of IW. This is a complex aspect that can be better understood when IW is compared with traditional forms of warfare. Traditionally, war entails the use of a state's *violence* through the state *military* forces to determine the conditions of governance over a determined territory (Gelven 1994). It is a necessarily violent phenomenon, which implies the sacrifice of human lives and damage to both military and civilian infrastructures. The problem to be faced when waging traditional warfare is how to minimise damage and losses while ensuring the enemy is overpowered.

IW is different from traditional warfare in several respects, mainly because it is not a necessarily violent and destructive phenomenon (Arquilla 1998; Dipert 2010; Barrett 2013). For example, IW may involve a computer virus capable of disrupting or denying access to the enemy's database, and in so doing it may cause severe damage to the opponent without exerting *physical* force or violence. In the same way, IW does not necessarily involve human beings. An action of war in this context can be conducted by an autonomous robot, such as, for example, the EADS Barracuda, and the Northrop Grumman X-47B[9], or by an autonomous cruising computer virus (Abiola et al. 2004), targeting other artificial agents or informational infrastructures, like a database or a website. IW can be waged exclusively in a digital context without ever involving physical targets, nevertheless it may escalate to more violent forms (Arquilla 2013; Waltz 1998; Clarke 2012; Brenner 2011; Bowden 2011).

[8] The USA only spent $ 400 million in developing technologies for cyber conflicts: http://www.wired.com/dangerroom/2010/05/cyberwar-cassandras-get-400-million-in-conflict-cash/.

The UK devoted £ 650 million to the same purpose: http://www.theinquirer.net/inquirer/news/1896098/british-military-spend-gbp650-million-cyber-warfare.

[9] Note that MQ-1 Predators and EADS Barracuda, and the Northrop Grumman X-47B are Unmanned Combat Aerial Vehicles used for combat actions and they are different from Unmanned Air Vehicles, like for example Northrop Grumman MQ-8 Fire Scout, which are used for patrolling and recognition purposes only.

Consider for example, the data diffused for GridExII.[10] This is a simulation that has been conducted in the US in November 2013. More than 200 utility companies collaborated with US government to simulate a massive cyber attack on US basic infrastructure. Had the attack been real, estimates mention hundreds of injuries and tens of deaths, while millions of US Citizens would have been left in the darkness.

As remarked above, the transversality of IW is the key feature of this phenomenon; it is the aspect that differentiates it the most from traditional warfare. Transversality is also the feature that engenders the ethical problems posed by IW. The potential bloodless and non-destructive nature of IW (Denning 2007; Arquilla 2013) makes it desirable from both an ethical and a political perspective, since at first glance, it seems to avoid bloodshed and it liberates political authority from the burden of justifying military actions to the public. However, the disruptive outcomes of IW can inflict serious damage to contemporary information societies at the same time, IW has the potential to lead to highly violent and destructive consequences, which would be dangerous for both military forces and civil society.

The need for strict regulations for declaring and waging IW in fairness is now compelling. To this end an analysis that discloses the ethical issues related to IW while pointing at the direction for their solution is a preliminary and necessary step. This will be the task of the next section.

## 7.3 IW and Just War Theory

Ethical analyses of war are developed following three main paradigms: JWT, Pacifism or Realism. In the rest of this paper, the analysis will focus only on JWT. Two reasons support this choice: the ethical problems with which JWT is concerned are generated by the very same decision to declare and to wage war, be it a traditional or an informational war. Therefore JWT sheds light on the analysis of the ethical issues posed by possible declaration of IW. More in general, the criteria for a *just* war proposed by this theory remain valid when considering IW, the justification to resort to war and the criteria for *jus in bello* and *post bellum* proposed are desirable also in case of IW and there is no doubt that just war principles and their preservation hold in the case of traditional warfare as well as in the case of IW.

Nevertheless, it would be mistaken to consider JWT both the necessary and sufficient ethical framework for the analysis of IW, for addressing this new form of warfare solely on the basis of JWT generates more ethical conundrums than it solves. The problem arises because JWT mainly focuses on the use of force in international contexts and surmises sanguinary and violent warfare occurring in the physical domain. As the cyber domain is virtual and IW mainly involves abstract entities, the application of JWT becomes less direct and intuitive. The struggle encountered when applying JWT to the cases of IW becomes even more evident if

[10] http://www.nytimes.com/2013/11/15/us/coast-to-coast-simulating-onslaught-against-power-grid.html.

one considers how pivotal concepts such as the ones of harm, target, attack have been reshaped by the dissemination of IW.[11] The very notion of harm for example, which is at the basis of JWT, struggle to apply to the case of. This a problem has been already highlighted in the extant literature, see for example (Dipert 2010) who argues that any moral analysis of this kind of warfare needs to be able to account for a notion of harm "[focusing] away from strictly injury to human beings and physical objects toward a notion of the (mal-) functioning of information systems, and the other systems (economic, communication, industrial production) that depend on them" (p. 386).

Particularly relevant to shed some light on the novelty posed by IW, is the transversality of the ontological status of the entities involved in the latter. Traditional warfare concerns human beings and physical objects, while IW involves artificial and non-physical entities alongside human beings and physical objects. Therefore, there is a *hiatus* between the ontology of the entities involved in traditional warfare and of those involved in IW. Such a hiatus affects the ethical analysis, for JWT rests on an anthropocentric ontology, i.e. it is concerned with respect for human rights and disregards all non-human entities as part of the moral discourse, and for this reason it does not provide sufficient means for addressing the case for IW (more details on this aspect presently).

The gap between the ontology assumed by JWT and the one of IW has also been described by Dipert, who stresses that "[s]ince cyber warfare is by its very nature information warfare, an ontology of cyber warfare would necessarily include way of specifying *information objects* [...], *the disruption and the corruption of data and the nature and the properties of malware*. This would be in addition to what would be required of a domain-neutral upper-level ontology, which addresses this type of characteristics of the most basic categories of entity that are used virtually in sciences and domain: material entity, event, quality of an object, physical object. A cyber warfare ontology would also go beyond [...] of a military ontology, such as agents, intentional actions, unintended effects, organizations, artefacts', commands, attacks and so on" (emphasis added) (Dipert 2013, p. 36).

The case of the autonomous cruising computer virus will help in clarifying the problems at stake (Abiola et al. 2004). These viruses are able to navigate through the web and identify autonomously their targets and attack them without requiring any supervision. The targets are chosen on the basis of parameters that the designers encode in the virus, so there is a boundary to the autonomy of these agents. Still, once the target has been identified the virus attacks without having to receive 'authorisation' from the designer or any human agent.

In considering the moral scenario in which the virus is launched three main questions arise. The first question revolves around the identification of the moral agents, for it is unclear whether the virus itself should be considered the moral agent, or

[11] The need to define concepts such as those of harm, target and violence is stressed both by scholar who argue in favor of the ontological difference of the cyber warfare (Dipert 2013) and exploit this point to claim that JWT is not an adequate framework to address IW and by those who actually maintain that JWT provides sufficient element to address the case of IW.

whether such a role should be attributed to the designer or to the agency that decided to deploy the virus, or even to the person who actually launched it. The second question focuses on moral patients. The issue arises as to whether the attacked computer system itself should be considered the moral receiver of the action, or whether the computer system and its users should be considered the moral patients. Finally, the third questions concerns the rights that should be defended in the case of a cyber attack. In this case, the problem is whether any rights should be attributed to the informational infrastructures or to the system compounded by the informational infrastructure and the users.

As noted by Dipert, IW includes informational infrastructures, computer systems, and databases. In doing so, it brings new objects, some of which are intangible, into the moral discourse. The first step toward an ethical analysis of IW is to determine the moral status of such (informational) objects and their rights. Help in this respect is provided by Information Ethics, which will be introduced in the Sect. 4. Before focusing on Information Ethics, we shall first consider in detail some of the problems encountered when applying JWT to IW.

### *7.3.1 The Tenets of JWT and IW*

Let me begin this section by stressing that the proposed analysis does not claim that JWT does not adequately respond to contemporary global politics or to new methods for waging violent warfare.[12] In the rest of this section I shall analyse the tenets of *last resort*, *more good than harm*, and *non-combatants immunity* to consider the problems that arise when these principles, which are desirable also in case of IW, are applied to the occurrences of a war in the cyber (non-physical) domain. I argue that the nexus of the ethical problems posed by IW rest on the ontological hiatus between IW and JWT, for the latter focuses on violent warfare, bloodshed and physical damages, and these aspects are peculiar of kinetic warfare but are not peculiar of IW.

The principle of 'war as last resort' prescribes that a state may resort to war only if it has exhausted all plausible, peaceful alternatives to resolve the conflict in question, in particular diplomatic negotiations. This principle rests on the assumption that war is a violent and sanguinary phenomenon and as such it has to be avoided until it remains the only reasonable way for a state to defend itself. The application of this principle is shaken when IW is taken in consideration, because in this case war may be bloodless and may not involve physical violence at all. In these circumstances, the use of the principle of war as last resort becomes less immediate.

Imagine, for example, the case of tense relations between two states and that the tension could be resolved if one of the states decide to launch a cyber attack on the other state's informational infrastructure. The attack would be bloodless as it would affect only the informational grid of the other state and there would be no casual-

[12] See (Withman 2013) for an analysis of validity of JWT with respect to contemporary violent warfare.

ties. The attack could also lead to resolution of the tension and avert the possibility of kinetic war in the foreseeable future. Nevertheless, according to JWT, the attack would be an act of war, and as such it is forbidden as a first strike move.

The impasse is quite dramatic, for if the state decides not to launch the cyber attack it will be probably forced to engage in a sanguinary war in the future, but if the state authorises the cyber attack it will breach the principle of war as last resort and commit an unethical action. This example is emblematic of the problems encountered in the attempt to establish ethical guidelines for IW. In this case, the main problem is due to the transversality of the modes of combat described in Sect. 2, which makes it difficult to define unequivocal ethical guidelines.

In the light of the principle of last resort, soft and non-violent cases of IW can be approved as means for avoiding traditional war (Perry 2006), as they can be considered a viable alternative to bloodshed, which may be justly endorsed to avoid traditional warfare (Bok 1999). At the same time, even the soft cases of IW have a disruptive purpose—disrupting the enemy's (informational) (Arquilla and Ronfeldt 1997; Arquilla 2013). Such a disruptive intent, even when it is not achieved through violent and sanguinary means, must be taken in consideration by any analysis aiming at providing ethical guidelines for IW.[13]

Another problem arises when considering the principle of 'more good than harm'. According to this principle, before declaring war a state must consider the *universal* goods expected to follow from the decision to wage war, against the *universal* evils expected to result, namely the casualties that the war is likely to determine. The state is justified in declaring war only when the goods are proportional to the evils. This is a fine balance, which is straightforwardly assessed in the case of traditional warfare, where evil is mainly considered in terms of casualties and physical damages that may result from a war. The equilibrium between the goods and the evils becomes more problematic to calculate when IW is taken into consideration.

As the reader may recall, IW is transversal with respect to the level of violence. If strictly applied to the non-violent instances of IW, the principle of more good than harm leads to problematic consequences. For it may be argued that, since IW can lead to the victory over the enemy without determining casualties, it is a kind of warfare (or at least the soft, non-violent instances of IW) that is always morally justified, as the good to be achieved will always be greater than the evil that could potentially be caused.

[13] It is worthwhile noticing that the problem engendered by the application of the principle of last resort to the soft-cases of IW may also be addressed by stressing that these cases do not fall within the scope of JWT as they may be considered cases of espionage rather than cases of war, and as such they do not represent a 'first strike' and the principle of last resort should not be applied to them. One consequence of this approach is that JWT would address war scenarios by focusing on traditional cases of warfare, such as physical attacks, and on the deployment of robotic weapons, disregarding the use of cyber attacks. This would be quite a problematic consequence because, despite the academic distinction between IW and traditional warfare, the two phenomena are actually not so distinct in reality. Robotic weapons fight on the battlefield side by side with human soldiers, and military strategies comprise both physical and cyber attacks. By disregarding cyber attacks, JWT would be able to address only partially contemporary warfare, while it should take into consideration the whole range of phenomena related to war waging in order to address the ethical issues posed by it (for a more in depth analysis of this aspect see (Taddeo 2012)).

Nonetheless, IW may result in unethical actions—destroying a database with rare and important historical information, for example. If the only criteria for the assessment of harm in warfare scenarios remain the consideration of the physical damage caused by war, then an unwelcome consequence follows, for all the non-violent cases of IW comply by default to this principle. Therefore, destroying a digital resource containing important records is deemed to be an ethical action tout court, as it does not constitute physical damage *per se*.

The problem that arose with the application of this principle to the case of IW does not concern the validity *in se* of the principle. It is rather the framework in which the principle has been provided that becomes problematic. In this case, it is not the prescription that the goods should be greater than the harm in order to justify the decision to conduct a war, but rather is the set of criteria endorsed to assess the good and the harm that shows its inadequacy when considering IW.

A similar problem arises when considering the principle of 'discrimination and non-combatant immunity'. This principle refers to a classic war scenario and aims at reducing bloodshed, prohibiting any form of violence against non-combatants, like civilians. It is part of the *jus in bello* criteria and states that soldiers can use their weapons to target exclusively those who are "engaged in harm" (Walzer 2006, p. 82). Casualties inflicted on non-combatants are excused only if they are a consequence of a non-deliberate act. This principle is of paramount importance, as it prevents massacres of individuals not actively involved in the conflict. Its correctness is not questionable yet its application is quite difficult in the context of IW.

In classic warfare, the distinction between combatants and non-combatants reflects the distinction between military and civil society. In the last century, the spread of terrorism and guerrilla warfare weakened the association between non-combatants and civilians. In the case of IW such association becomes even feebler, due to the blurring between civil society and military organisations (Schmitt 1999; Shulman 1999; Taddeo 2012).

The blurring of the distinction between military and civil society leads to the involvement of civilians in war actions and raises a problem concerning the discrimination itself: in the IW scenario it is difficult to distinguish combatants from non-combatants. Wearing a uniform or being deployed on the battlefield are no longer sufficient criteria to identify someone's social status. Civilians may take part in a combat action from the comfort of their homes, while carrying on with their civilian life and hiding their status as informational warriors.

This case provides also a good example of the policy gap surrounding IW, for one of the most important aspects of the distinction between military and civilian concerns the identification of the so-called civilian objects, i.e. buildings, places and objects that should not be considered military targets. Chapter III of the Protocol I of the Geneva Convention[14] defines civilian objects as material tokens, which are further categorised of cultural or religious type, environmental or necessary to the survival of the population. This chapter shows to be ontologically limited as it considers as 'objects' only physical, tangible entities.[15]

[14] "ICRC Databases on International Humanitarian Law".

[15] On this point see also (Dipert 2010, p. 400).

Furthermore, civilian objects are distinguished from military one, as the latter are deemed to be objects that "make an effective contribution to military action and whose total or partial destruction, capture or neutralization, in the circumstances ruling at the time, offers a definite military advantage". The reader may easily see how such a definition may be used to qualify a civilian informational infrastructure in time of IW, making the line between civilian and military even less evident and making even more compelling the need for policies able to accommodate a more inclusive definition of objects, and more in general able to address the conceptual changes posed by this new kind of warfare.

Before introducing Information Ethics, I shall remark that several analyses have been proposed claiming that the existing apparatus of laws resting on JWT is adequate and sufficient to address the cases of IW. This is an interesting and also useful approach, it allows for applying current international laws to IW and to these days it avoided that the cyber sphere would become an unregulated domain. However, the approach encounters a major flaw, for it equates JWT with the body of international and national laws regulating warfare and overlooks the conceptual roots on which this theory rests. In doing so, the universal nature of JWT is missed and so is the possibility of expanding the scope of this theory by reshaping its conceptual framework. The consequence is that rather than revising the conceptual roots of JWT in order to address the novelty posed by IW, the latter is 'forced' to fit the parameters set for kinetic warfare.

The approach hence fails to consider and to account for the conceptual changes prompted by IW (see the ones discussed in Sect. 2 and 3) and risks to confuse the remedy for the solution and, in the long run, to pose conceptual limitations to the laws and regulation for IW. A good example in this respect concerns application of the principle of just cause to IW. As Barrett (2013) put it "[s]ince damage to property may constitute a just cause, can temporary losses of computer functionality also qualify as a *casus belli*? Like kinetic weapons, cyber-weapons can physically destroy or damage computers. But offensive computer operations, because of their potential to be transitory or reversible, can also merely compromise functionality. While permanent loss of functionality create the same effect as physical destruction, temporary functionality losses are unique to cyber-operations and require additional analysis" (p. 6).

The issue is not whether the case of IW can be considered in such a way to fit the parameters of kinetic warfare and hence to fall within the domain of JWT, as we know it. This result is easily achieved once the focus is restricted to physical damage and tangible objects. The problem lays at a deeper level and questions the very conceptual framework on which JWT rests and it ability to *satisfactory* and *fairly* accommodate the changes brought to the fore by the information revolution, which are affecting the way we wage war, but also the way in which we conduct our lives, perceive ourselves and the very concepts of harm, warfare, property, state.

It would be misleading to consider the problems described in this sections as reasons for dismissing JWT when analysing IW. These problems rather point to a more fundamental issue; namely the need to consider more carefully the case of IW, and to take into account its peculiarities.

## 7.4 Information Ethics

Information Ethics is a macro-ethics, which is concerned with the whole realm of reality and provides an analysis of ethical issues by endorsing an informational perspective. Such an approach rests on the consideration that "ICTs, by radically changing the informational context in which moral issues arise, not only add interesting new dimensions to old problems, but lead us to rethink, methodologically, the very grounds on which our ethical positions are based" (Floridi 2006, p. 23).

In one sentence Information Ethics is defined as a *patient-oriented, ontocentric*, and *ecological* macroethics. Information Ethics is patient-oriented because it considers the morality of an action with respect to its effects on the receiver of the action. It is ontocentric, for it endorses a non-anthropocentric approach for the ethical analysis. It attributes a moral value to all the existing entities (both physical and non-physical) by applying the principle of ontological equality: "This ontological equality principle means that any form of reality [...], simply for the fact of being what it is, enjoys a minimal, initial, *overridable*, equal right to exist and develop in a way which is appropriate to its nature" (Floridi Forthcoming). The principle of ontological equality is grounded on an information-based ontology,[16] according to which all existing things can be considered from an informational standpoint and are understood as informational entities, all sharing the same informational nature.

The principle of ontological equality shifts the standing point for the assessment of the moral value of entities, including technological artefacts. At first glance, an artefact, a computer, a book or the Colosseum, seems to enjoy only an instrumental value. This is because in considering them one endorses an anthropocentric LoA, in other words one considers these objects as a user, a reader, a tourist. In all these cases the moral value of the observed entities depends on the agent interacting with them and on her purpose in doing so.

The claim put forward by Information Ethics is that, these LoAs are not adequate to support an effective analysis of the moral scenario in which the artefacts may be involved. The anthropocentric, or even the biocentric, LoA prevent to properly consider the nature and the role of such artefacts in the reality in which we live. The argument is suggested that all existing things have an informational nature, which is shared across the entire spectrum—from abstract to physical and tangible entities, from rocks and books to robots and human beings, and that all entities enjoy some minimal initial moral value *qua informational* entities.

Information Ethics argues that universal moral analyses can be developed by focusing on the common nature of all existing things and by defining good and evil with respect to such a nature. The focus of the ethical analysis is shifted, the initial moral value of an entity does not depend on the observer, but is defined in absolute terms and depends on the (informational) nature of the entities. Following the principle of ontological equality, minimal and overridable rights to exist and flourish

[16] The reader may recall the informational LoA mentioned in Sect. 2. Information Ethics endorses an informational LoA, as such it focuses on the informational nature as a common ground of all existing things.

pertain to all existing things and not just to human or living things. The Colosseum, Jane Austin's writings, a human being and computer software all share initial right to exist and flourish, as they are all informational entities.[17]

A clarification is now necessary. Information Ethics endorses a minimalist approach, it considers informational nature as the minimal common denominator among all existing things. Such a minimalist approach should not be mistaken for reductionism, as Information Ethics does not claim that the informational one is the unique LoA from which moral discourse is addressed. Rather it maintains that the informational LoA provides a *minimal starting point*, which can then be enriched by considering other moral perspectives.

Lest the reader be mislead, it is worthwhile emphasising that the principle of ontological equality does not imply that all entities have the same moral value. The rights attributed to the entities are *initial*, they are overridden whenever they conflict with the rights of other (more morally valuable) entities. Furthermore, the moral value of an entity is determined according to its potential contribution to the enrichment and the flourishing of the informational environment. Such an environment, the *Infosphere*, includes all existing things, be they digital or analogical, physical or non-physical and the relations occurring among them, and between them and the environment. The blooming of the Infosphere is the ultimate good, while its corruption, or destruction, is the ultimate evil.

In particular, any form of corruption, depletion and destruction of informational entities or of the Infosphere is referred to as *entropy*. In this case entropy refers to "any kind of *destruction* or *corruption* of informational objects (mind, not of information), that is, any form of impoverishment of *being*, including *nothingness*, to phrase it more metaphysically", (Floridi Forthcoming) and has nothing to do with the concept developed in physics or in information theory (Floridi 2007).

Information Ethics considers the duty of any moral agent with respect to its contribution to the informational environment, and considers any action that affects the environment by corrupting or damaging it, or by damaging the informational objects existing in it, as an occurrence of entropy, and therefore as an instance of evil (Floridi and Sanders 2001). On the basis of this approach Information Ethics provides four principles to identify right and wrong and the moral duties of an agent. The four moral principles are:

1. entropy ought not to be caused in the infosphere (null law);
2. entropy ought to be prevented in the infosphere;
3. entropy ought to be removed from the infosphere;
4. the flourishing of informational entities as well as of the whole infosphere ought to be promoted by preserving, cultivating and enriching their properties.

These four principles together with the theoretical framework of Information Ethics will provide the ground to proceed further in our analysis, and define the principles for a just IW.

[17] For more details on the information-based ontology see (Floridi 2002). The reader interested in the debate on the Informational ontology and the principles of Information Ethics may whish to see (Floridi 2007).

## 7.5 Just IW

The first step toward the definition of the principles for a just IW is to understand the moral scenario determined by this phenomenon. The framework provided by Information Ethics proves to be useful in this regard, for we can now answer the questions posed in Sect. 3 concerning the identification of moral agents, moral patients and the rights that have to be respected in the case of IW. The remainder of this chapter will not focus on the problems regarding moral patients and their rights. The issue concerning the identification of moral agents in IW requires an in-depth analysis (see for example (Asaro)) which falls outside the scope of this chapter. I shall clarify a few aspects concerning morality of artificial agents relevant to the scope of this analysis, before setting this issue aside.

The debate on morality of artificial agents is usually associated to the issues of ascribing to artificial agents moral responsibility for their actions. Floridi and Sanders (2004) provide a different approach to this problem decoupling the moral *accountability* of an artificial agent, i.e. its ability to perform morally qualifiable actions, from the moral *responsibility* for the actions that such an agent may perform.

The authors argue that an action is morally qualifiable when it has morally qualifiable effects, and that every entity that qualifies as an interactive, autonomous and adaptable (transition) system and which performs a morally qualifiable action is (independently from its ontological nature) considered a morally accountable agent. So when considering the case for IW, a robotic weapon and a computer virus are considered moral agents as long as they show some degree of autonomy in interacting and adapting to the environment and perform actions that may cause either moral good or moral evil.

As argued by Floridi and Sanders, attributing moral accountability to artificial agents extends the scope of ethical analysis to include the actions performed by the agents and permits to determine moral principles to regulate such actions. This approach particularly suits the purpose of the present analysis, for the reader may accept suspending judgment on the moral responsibility for the actions that artificial agents may perform in case of IW, and agree that such actions are nevertheless morally qualifiable, and that as such they should be the objects of a prescriptive analysis.

Once we have put aside the issue concerning the morality of artificial agents, we are left with questions concerning the moral stance of the receivers of the actions performed by such agents and of the rights that ought to be respected in the case of IW. The principle of ontological equality states that all (informational) entities enjoy some minimal initial rights to exist and flourish in the Infosphere, and therefore every entity deserves some minimal respect, in the sense of a "disinterested, appreciative and careful attention" (Hepburn 1984; Floridi 2013).

When applied to IW, this principle allows for considering all entities that may be affected by an action of war as moral patients. A human being, who enjoys the consequences of a cyber attack and an informational infrastructure that is disrupted by a cyber attack are both to be held moral patients, as they are both the receivers of the moral action. Following Information Ethics, the moral value of such an action is to be assessed on the basis of its effects on the patients' rights to exist and flourish, and ultimately on the flourishing of the Infosphere.

The issue then arises concerning which and whose rights should be preserved in case of IW. The answer to this question follows from the rationale of Information Ethics, according to which an entity may lose its rights to exist and flourish when it comes into conflict (causes entropy) with the rights of other entities or with the well-being of the Infosphere. It is a moral duty of the other inhabitants of the Infosphere to *remove* such a malicious entity from the environment or at least to impede it from perpetrating more evil.

This framework lays the ground for the first principle for just IW. The principle prescribes the condition under which the choice to resort to IW is morally justified.

I. IW ought to be waged only against those entities that endanger or disrupt the well-being of the Infosphere.

Two more principles regulate just IW, they are:

II. IW ought to be waged to preserve the well-being of the Infosphere.
III. IW ought not to be waged to promote the well-being of the Infosphere.

The second principle limits the task of IW to restoring the *status quo* in the Infosphere before the malicious entity began increasing the entropy within it. IW is just as long its goal is to *repair* the Infosphere from the damage caused by the malicious entity.

The second principle can be described using an analogy; namely, IW should fulfil the same role as police forces in a democratic state. It should act only when a crime has been, or is about to be, perpetrated. Police forces do not act in order to ameliorate the aesthetics of cities or the fairness of a state's laws; they only focus on reducing or preventing crimes from being committed. Likewise, IW ought to be endorsed as an *active* measure in response to increasing of evil and not as proactive strategy to foster the flourishing of the Infosphere. Indeed, this is explicitly forbidden by the third principle, which prescribes the promotion of the well-being of the Infosphere as an activity that falls beyond the scope of a just IW.

These three principles rest on the identification of the moral good with the flourishing of the Infosphere and the moral evil with the increasing of entropy in it. They endorse an informational ontology, which allows for including in the moral discourse both non-living and non-physical entities. The principles also prescribe respect for the (minimal and overridable) rights of such entities along with those of human beings and other living things, and respect for the rights of the Infosphere as the most fundamental requirement for declaring and waging a just IW.

In doing so the three principles overcome the ontological hiatus described in Sect. 3, and provide the framework for applying JWT to the case of IW without leading to the ethical conundrums analysed in Sect. 3.1. The description of how JWT is merged with Information Ethics is the task of the next section.

## 7.6 Three Principles for a Just IW

The application of the principle of 'last resort' provides the first instance of the merging of JWT and Information Ethics. The reader may recall that the principles forbids embracing IW as an 'early move' even in those circumstances in which IW may avert the possibility of waging a traditional war. The principle takes into account traditional (violent) forms of warfare, and it is coupled with the principle of 'right cause', which justifies resort to war only in case of 'self-defence'. However right this approach may be when applied to traditional (violent) forms of warfare, it proves inadequate when IW is taken into consideration. The impasse is overcome when considering the principles for just IW.

The first principle prescribes that any entity that endangers or disrupts the well-being of the Infosphere loses its basic rights and becomes a licit target. The second principle prescribes that a state is within its rights to wage IW to re-establish the *status quo* in the Infosphere and to repair the damage caused by a malicious entity. These two principles allow for breaking the deadlock described in Sect. 3.1, because a state can rightly endorse IW as an early move to avoid the possibility of a traditional warfare, as the latter threatens greater disruption of the Infosphere, and as such it is deemed to be a greater evil (source of entropy) than IW.

A caveat must be stressed in this case: the waging of IW must comply with the principles of 'proportionality' and 'more good than harm'. In waging IW, the endorsed means must be sufficient to stop the malicious entity, and in doing so the means ought not to generate more entropy than a state is aiming to remove from the Infosphere in the first place. This leads us to consider in more detail the principle of 'more good than harm'.

The issues that arose in the case of IW are due to the definition of the criteria for the assessment of the 'good' and the 'harm' that a warfare may cause. As described in Sect. 3.1, endorsing traditional criteria leads to a serious ethical conundrum, since all (the majority of) the cases of IW that do not target physical infrastructures or human life comply by default to this principle regardless of their consequences.

Such a problem is avoided if damage to non-physical entities in considered as well as physical damage. More precisely, the assessment of the good and the harm should be determined by considering the general condition of the Infosphere 'before and after' waging the war. A just war never determines greater entropy than that in the Infosphere before it was waged. Once considered from this perspective, the principle of more good than harm acts as corollary of the second principle for just IW. It ensures that a just IW is waged to restore the *status quo* and does not increase the level of entropy in the Infosphere.

Increasing entropy in the Infosphere also provides a criterion for reconsidering the application of the principle of 'discrimination and non-combatants' immunity' to IW. As it has been argued in Sect. 3.1, IW blurs the distinction between militaries and civilians, as it neither requires military skills nor does it require a military status of the combatants to be waged. This makes problematic the application of this principle to IW; nevertheless the principle has to be maintained as it prescribes the distinction between licit and illicit war targets.

Help in applying this principle to IW comes from the first principle for just IW, which allows for dispensing with the distinction between militaries and civilians, and for substituting it with the distinction between licit targets and illicit ones. The former are those malicious entities who endanger or disrupt the well-being of the Infosphere. According to the principle, IW rightfully targets only malicious entities, be they military or civilian. The social status ceases to be significant in this context, because any entity that contributes to increasing the evil in the Infosphere loses its initial rights to exist and flourish and therefore becomes a licit target. More explicitly, it becomes a moral duty for the other entities in the Infosphere to prevent such entity from causing more evil.

Before concluding this chapter, I shall briefly clarify an aspect of the proposed analysis, lest the reader be tempted to consider it warmongering.

The third principle provided in Sect. 5 stresses that IW is never justly waged when the goal is improving the well-being of the Infosphere. This principle rests on the very same rationale that inspires Information Ethics, according to which the flourishing of the Infosphere is determined by the blooming of informational entities, of their relations and by their well-being. IW is understood as a form of disruption and as such, by definition, it can never be a vehicle for fostering the prosperity of the Infosphere nor is it deemed to be desirable *per se*. IW is rather considered a necessary evil, the bitter medicine, which one needs to take to fight something even more undesirable, i.e. the uncontrolled increasing of the entropy in the environment. With this clarification in mind we can now pull together the threads of the analysis proposed in this chapter.

## 7.7 Conclusion

The goal of this chapter is to fill the conceptual vacuum surrounding IW and of providing the ethical principles for a just IW. It has been argued that to this purpose JWT provides the necessary but not sufficient tools. For, although its ideal of just warfare grounded on respect for basic human rights in the theatre of war holds also in the case of IW, it does not take into account the moral stance of non-human and non-physical entities which are involved and mainly affected by IW. This is the ontological hiatus, which I identified as the nexus of the ethical problems encountered by IW.

This chapter defends the thesis that in order to be applied to the case for IW, JWT needs to extend the scope of the moral scenario to include non-physical and non-human agents and patients. Information Ethics has been introduced as a suitable ethical framework capable of considering human and artificial, physical and non-physical entities in the moral discourse. It has been argued that the ethical analysis of IW is possible when JWT is merged with Information Ethics. In other words, JWT *per se* is too large a sieve to filter the issues posed by IW. Yet, when combined with Information Ethics, JWT acquires the necessary granularity to address the issues posed by this form of warfare.

The first part of this paper introduces IW and analyses its relation to the information revolution and its main feature, namely its transversality. It then describes the reasons why JWT is an insufficient tool with which to address the ethical problems engendered by IW and continues by introducing Information Ethics. The second part of the chapter defends the thesis according to which once the ontological hiatus between the JWT and IW it is bridged, JWT can be endorsed to address the ethical problems posed by IW.

The argument is made that such a hiatus is filled when JWT encounters Information Ethics, since its ontocentric approach and informational ontology allow for ascribing a moral status to any existing entity. In doing so, Information Ethics extends the scope of the moral discourse to all entities involved in IW and provides a new ground for JWT, allowing it to be extended to the case for IW.

In concluding this chapter I should like to remark that the proposed ethical analysis should in no way be understood as a way of advocating warfare or IW. Rather it is devoted to prescribing ethical principles such that if IW has to be waged then it will at least be a just warfare.

## References


Abiola, A., J. Munoz, and W. Buchanan. 2004. Analysis and detection of cruising computer viruses. In Proceedings of 3rd *EIWC*.

Arquilla, J. 1998. Can information warfare ever be just? *Ethics and Information Technology* 1 (3): 203–212.

Arquilla, J. 2013. Twenty years of cyberwar. *Journal of Military Ethics* 12 (1): 80–87. doi:10.1080/15027570.2013.782632.

Arquilla, J., and D. F. Ronfeldt, eds. 1997. *In Athena's camp: Preparing for conflict in the information age*. Santa Monica: Rand.

Asaro, P. 2008. How just could a robot war be? In: Brey P, Briggle A, Waelbers K (eds) Current issues in computing and philosophy. IOS Press, Amsterdam, pp 50–64.

Barrett, E. T. 2013. Warfare in a new domain: The ethics of military cyber-operations. *Journal of Military Ethics* 12 (1): 4–17. doi:10.1080/15027570.2013.782633.

Benbow, T. 2004. *The magic bullet?: Understanding the "revolution in military affairs"*. London: Brassey's.

Blackmore, T. 2011. *War X*. Toronto: University of Toronto Press.

Bok, S. 1999. *Lying: Moral choice in public and private life*. 2nd ed. New York: Vintage Books.

Bowden, M. 2011. *Worm: The first digital world war*. New York: Atlantic Monthly Press.

Brenner, J. 2011. *America the vulnerable: New technology and the next threat to national security*. New York: Penguin Press.

Clarke, R. A. 2012. *Cyber war: The next threat to national security and what to do about it*. 1st ed. New York: Ecco.

Denning, D. 2007. The ethics of cyber conflict. In *Information and computer ethics*, eds. K. E. Himma and H. T. Tavani. Hoboken, USA: Wiley.

Dipert, R. 2010. The ethics of cyberwarfare. *Journal of Military Ethics* 9 (4): 384–410.

Dipert, R. 2013. The essential features of an ontology for cyberwarfare. In *Conflict and cooperation in cyberspace,* eds. Panayotis Yannakogeorgos and Adam Lowther, 35–48. NY: Taylor & Francis. http://www.crcnetbase.com/doi/abs/10.1201/b15253-7. Accessed on 12 Oct 2013.
Floridi, L. 2002. On the intrinsic value of information objects and the infosphere. *Ethics and Information Technology* 4 (4): 287–304.
Floridi, L. 2006. Information ethics, its nature and scope. SIGCAS Comput. Soc., 36(3): 21–36.
Floridi, L. 2007. Understanding information ethics. *APA Newsletter on Philosophy and Computers* 7 (1): 3–12.
Floridi, L. 2008. The method of levels of abstraction. *Minds and Machines* 18 (3): 303–329. doi:10.1007/s11023-008-9113-7.
Floridi, L. 2013. *Ethics of information.* Oxford: Oxford University Press.
Floridi, L. 2014. *The fourth revolution, how the infosphere is reshaping human reality.* Oxford: Oxford University Press.
Floridi, L., and J. Sanders. 2001. Artificial evil and the foundation of computer ethics. *Ethics and Information Technology* 3 (1): 55–66.
Floridi, L., and J. W. Sanders. 2004. On the morality of artificial agents. *Minds and Machines* 14 (3): 349–379. doi:10.1023/B:MIND.0000035461.63578.9d.
Gelven, M. 1994. *War and existence: A philosophical inquiry.* University Park: Pennsylvania State University Press.
Hepburn, R. W. 1984. *"Wonder" and other essays: Eight studies in aesthetics and neighbouring fields.* Edinburgh: University Press.
"ICRC Databases on International Humanitarian Law." 00:00:00.0. http://www.icrc.org/ihl/INTRO/470. Accessed on 12 Oct 2013.
Libicki, M. 1996. *What is information warfare?* Washington, D.C.: National Defense University Press.
Perry, D. 2006. 'Repugnant philosophy': Ethics, espionage, and covert action. In *Ethics of spying: A reader for the intelligence professional,* ed. J. Goldman. Scarecrow press, Oxford: UK, p. 221–247.
Schmitt, M. N. 1999. The principle of discrimination in 21st century warfare. SSRN Scholarly Paper ID 1600631. Rochester: Social Science Research Network. http://papers.ssrn.com/abstract=1600631. Accessed on 12 Oct 2013.
Schwartau, W. 1994. *Information warfare: Chaos on the electronic superhighway.* 1st ed. New York: Thunder's Mouth Press. (Distributed by Publishers Group West).
Shulman, M. R. 1999. Discrimination in the laws of information warfare. SSRN Scholarly Paper ID 1287181. Rochester, NY: Social Science Research Network. http://papers.ssrn.com/abstract=1287181. Accessed on 12 Oct 2013.
Taddeo, M. 2012. Information warfare: A philosophical perspective. *Philosophy & Technology* 25 (1): 105–120.
Taddeo, M., and A. Vaccaro. 2011. Analyzing peer-to-peer technology using information ethics. *The Information Society* 27 (2): 105–112. doi:10.1080/01972243.2011.548698.
Toffler, A., and H. Toffler. 1997. Foreword: The new intangibles. In *In Athena's camp preparing for conflict in the information age,* eds. John Arquilla and David F. Ronfeldt. Santa Monica: Rand.
Turilli, M., A. Vaccaro, and M. Taddeo. 2011. Internet neutrality: Ethical issues in the internet environment. *Philosophy & Technology* 25 (2): 133–151. doi:10.1007/s13347-011-0039-2.
Waltz, E. 1998. *Information warfare: Principles and operations.* Boston: Artech House.
Walzer, M. 2006. *Just and unjust wars: A moral argument with historical illustrations.* 4th ed. New York: Basic Books.
Withman, J. 2013. Is just war theory obsolete? In *Routledge handbook of ethics and war: Just war theory in the 21st century,* eds. Fritz Allhoff, Nicholas G. Evans, and Adam Henschke, Abimdon, UK: Routledge 23–34.

# Index



L. Floridi (ed.), *Protection of Information and the Right to Privacy – A New Equilibrium?*, Law, Governance and Technology Series 17, DOI 10.1007/978-3-319-05720-0,